Astronomers' Observing Guides

Other Titles in This Series

Forthcoming Title in This Series

Martin Mobberley

Total Solar Eclipses
and How to
Observe Them

with 115 Illustrations

Springer

Martin Mobberley
martin.mobberley@btinternet.com

Series Editor:
Dr. Mike Inglis, BSc, MSc, Ph.D.
Fellow of the Royal Astronomical Society
Suffolk County Community College
New York, USA
inglism@sunysuffolk.edu

Cover illustration:
Miloslav Druckmüller and Peter Aniol.
2006 March 29th Total Solar Eclipse.

Library of Congress Control Number: 2007923647

ISBN-13: 978-0-387-69827-4   e-ISBN-13: 978-0-387-69828-1

Printed on acid-free paper.

9 8 7 6 5 4 3 2 1

springer.com

Dedicated to the memory of Barbara Mobberley (1929–2006). For forty-eight years, a remarkably devoted mother, who always encouraged me in all my astronomical endeavours, and always worried about me when I was away on eclipse trips. She passed away a few hours after the crystal clear UK moonrise lunar eclipse of September 2006.

Preface

To children the world is full of magical events, and the line between make-believe and reality is (happily) distinctly blurred. However, as we all age and have to be realistic, earn money, become serious and responsible adults (yawn!) and accept that life is short and no one is immortal, those magical events fade and die. We accept that the worlds of make-believe are simply a product of the vivid imaginations of great story tellers, and that we are all very similar human beings; just slogging away at the tedious day job, and hoping for a win on the lottery. However, there are still a few events which are truly magical and, for me, total solar eclipses (TSEs) are about as magical an event as you can ever experience. Our Earth and Sun are, by a ludicrously lucky coincidence (or, some would argue, by a cosmic architect) almost the same apparent size in the sky. Thus, seventy times a century, the Moon's shadow passes over a narrow band on the Earth's surface and, for those on the track, with clear skies, a truly awe-inspiring sight can be seen.

But be warned, eclipse chasing is an addictive drug. Once you have seen totality, you will crave more of the same. The sight of an eclipsed Sun is the stuff of science fiction, but, for the cost of a plane flight or an organized holiday, it can be yours, albeit for a few fleeting minutes.

This is not the first total solar eclipse book to be written. A handful of others have been published in the last ten years. However, this is the first new total solar eclipse book to be published in the exciting digital photography era and so I make no apology for bringing the subject bang up to date.

Martin Mobberley
Suffolk, UK
January 2007

Acknowledgements

I am, as always, indebted to all those who have helped me with free images and information. Without such generosity very few of the books aimed at amateur astronomers would be financially viable. All of us in the eclipse-chasing community are permanently indebted to Fred Espenak ("Mr. Eclipse") and NASA who provide so much eclipse track data for us all. In addition, the weather prospects compiled by Jay Anderson are an indispensable aid to individual travellers and travel companies alike. I am especially grateful to Heinz Scsibrany for his excellent WinEclipse freeware and his permission to use it to produce many of the diagrams in this book. My special thanks are also due to Michael Gill who moderates the Solar Eclipse Mailing List (SEML) and generously provided a number of photographs and much anecdotal information. In addition I am indebted to Miloslav Druckmüller, the king of coronal image processing, who has generously made his exquisite images available for this book.

In alphabetical order I give my sincere thanks to the following contributors to this book: Jay Anderson, Martin Andrews, Peter Aniol, Chris Baddiley, Jamie Cooper, Miloslav Druckmüller and Hana Druckmüllerová, Fred Espenak, Patricia Totten Espenak, Ray Emery, Nigel Evans, David Hardy, Mike Harlow, Michael Gill, Pete Lawrence, Paul Maley, Eva Markova (Upice Observatory), Lynn Palmer, Jay Pasachoff, Damian Peach, Bob & Maria Priest, Vojtech Rusin, Glenn Schneider, Heinz Scsibrany, Eric Strach, Dave Tyler, Sheridan Williams and, last but not least, Val and Andrew White.

I would also like to thank my father for his constant support in all my observing projects.

Martin Mobberley
Suffolk, UK
January 2007

Contents

Part I Eclipse Mechanisms, Statistics and Tracks

Part II Observing and Travelling to Total Solar Eclipses

Contents

Part I

Eclipse Mechanisms, Statistics and Tracks

Chapter 1

Why Do Eclipses Occur?

A Remarkable Moon

We live on a planet that, for its size, has an unusually large Moon. Indeed, before the discovery of Pluto's Moon, Charon, in 1978, the relative size of our satellite was unique in the solar system. Moving outwards from the Sun and therefore dealing with the terrestrial planets first: Mercury and Venus have no moons, Earth has one, Mars has two tiny moons and then we arrive at the asteroid belt. On the other side of this swarm of hundreds of thousands of minor planets we encounter the four gas giants. Jupiter has three moons that are larger than our own, namely Io, Ganymede and Callisto; Europa is only slightly smaller, but they are dwarfed by the 140,000 km diameter parent planet. Saturn has an incredible 30 moons, but Titan, the only moon in our solar system with an atmosphere, is the only one larger than Earth's Moon. An image of Saturn itself, eclipsing the Sun with ease, is shown in Fig. 1.1. None of the 20 satellites of Uranus come anywhere near the size of our Moon and distant Neptune's large satellite Triton is only three-quarters our own Moon's diameter. It is really astonishing that the Earth has such a relatively large satellite (Table 1.1).

But, it is in terms of relative size that the rarity of Earth's Moon really hits home. Our Moon has an equatorial diameter of 3,476 km, compared to the Earth's 12,756 km; it is 27% of the Earth's diameter and 1/81 of the Earth's mass. Compare these ratios with those of Jupiter and its huge moon Ganymede. Jupiter has an equatorial diameter of 142,884 km compared with 5,268 km for Ganymede. That moon is therefore only 3.7% of the parent planet's diameter and 1/12,700 of its mass. Most remarkable of all is the fact that our Moon is 400 times closer to us than the Sun, and it is 400 times smaller too. These two factors mean that we can experience near-perfect solar eclipses, where the blinding solar disc is just covered, but the Sun's corona and prominences are fully revealed. If we really look hard enough in our Solar System we can find other cases where a satellite can perfectly eclipse the Sun. I am specifically thinking here of an astronaut standing on one of the Jovian Moons around the far side of Jupiter looking towards another Moon on the Sunward side of its orbit and passing in front of the Sun. The Sun is only 6 arcmin across as seen from Jupiter, so the spectacle would be very tame compared with an eclipse of our Moon seen from the Earth. Nevertheless, on rare occasions, when the plane of the Jovian satellites line up with the ecliptic (every 6 years) such eclipses can take place. Indeed, they can even be observed from Earth, as the shadow of one Galilean satellite on the earthward side of its orbit eclipses another on the farthest part of its orbit. Io can eclipse the Sun as seen from Ganymede or

Fig. 1.1. Other planets can be made to eclipse the Sun, especially if you are onboard a spacecraft. The Cassini probe took this emerging Diamond Ring image of the Sun behind Saturn on 15 September 2006. Image: Nasa/JPL

Table 1.1. The solar system's largest moons, plus Pluto's Charon (the largest relative to its parent) and how big they and the Sun appear when viewed from their home planet

Satellite	Diameter (km)	Parent planet	Rel. diam. w.r.t. parent (%)	Sun diam. (arcmin)	Moon diam. (arcmin)
Moon	3,476	Earth	27	31.5′–32.5′	29.4′–33.5′
Io	3,640	Jupiter	2.5	6′	36′
Europa	3,130	Jupiter	2.2	6′	18′
Ganymede	5,268	Jupiter	3.7	6′	18′
Callisto	4,806	Jupiter	3.4	6′	9′
Titan	5,150	Saturn	4.3	3.5′	15′
Triton	2,705	Neptune	5.4	1′	28′
Charon	1,190	Pluto	50	0.8′	4°

The angular size in arcminutes of the Sun and moon are shown, assuming an observer is positioned on the planet's equator (a gaseous "surface" for Jupiter, Saturn and Neptune). In the case of our Moon, the geocentric range of Sun and Moon sizes are shown. Note how small the big four Galilean Moons of Jupiter are with respect to Jupiter, i.e. between 2.2 and 3.7% of its diameter. Not only is our Moon unusually large compared to the parent planet, it has roughly the same angular size in the sky as the Sun, when viewed from the Earth, i.e. half a degree.

Europa; Europa can almost perfectly eclipse the Sun as seen from Ganymede; and Ganymede can eclipse the Sun as seen from Callisto. However interesting these facts are, the total solar eclipses (TSEs is a popular abbreviation) seen by intelligent eclipse chasers on our beautiful Earth and caused by our Moon are the showpiece eclipses in our solar system and, who knows, maybe in our galaxy too? As an aside, it is worth mentioning that the astronauts on the Apollo 14 mission just missed out on witnessing the Earth totally eclipse the Sun, as viewed from the Moon. They splashed down in the Pacific Ocean at 21.05 G.M.T. on 9 February 1971, 10 h before a total lunar eclipse was visible from Earth, and a total solar eclipse, by the Earth, would have been visible from their Fra Mauro landing site. Of course, the Earth would have appeared almost 2° across, i.e. nearly four times the size of the Sun, and therefore engulfing almost all the solar corona at maximum eclipse.

Incidentally, I have deliberately left lunar eclipses out of this book simply because, while they are beautiful events, they are not mind-blowing spectacles that make grown men cry. The Full Moon enters the Earth's shadow and glows bright red or dark red as light is bent through Earth's atmosphere onto the Moon. The Earth's umbral shadow is, on average, 2.7 times our Moon's diameter, at the Moon's distance, and so the Moon is totally swallowed up by it. Perhaps surprisingly a lunar eclipse is rarer than an eclipse of the Sun; however, the critical point is that everyone on the night side of the Earth can see them, not just an elite band of people on a small track.

The Origin of the Moon

So how come Earth has such a huge satellite? We would expect Jupiter, with its colossal gravitational influence, to capture a lot of nearby passing debris in the formative years of the solar system, but Earth is just too small to achieve this. Take a look at our fellow inner solar system planets, namely Mercury, Venus and Mars; they have failed to grab anything substantial and the tiny Martian moons of Phobos and Deimos can hardly be considered as anything more than debris. The mystery of our Moon's origin has long puzzled astronomers and geologists, but it was only with the advent of modern computers and the lunar samples returned from the Apollo missions that the picture became clearer. The old and naïve idea that our Moon somehow emerged as a molten blob from the Earth's primordial sphere (from the region we now call the Pacific Ocean according to some old books) has long been consigned to history. Astronomers now believe that, billions of years ago, the body that would become the Earth was in collision with another planet-sized object, maybe similar in size to Mars (but, I should stress, *not* Mars). Two crucial papers published in 1975, by William K. Hartmann and Donald Davis and another by Alfred G.W. Cameron and William Ward independently laid the groundwork for the current consensus (almost) for the impact scenario. The Hartmann/Davis argument was largely based on the enormous evidence for huge impacts forming the largest lunar craters and basins: If 100–200 km impactors were around in the early solar system, why not bigger objects and planet-sized collisions? The Cameron/Ward argument was more mathematical and centred around the amount of angular momentum in the Earth–Moon system, i.e. the rotation of both bodies and the motion of the Moon around the Earth. Cameron's estimate was that an object roughly 10% of the Earth's mass (i.e. a bit smaller than Mars' size) would have been responsible for side-swiping (or even double impacting) the Earth and creating the ring of debris that eventually formed the molten Moon, possibly in only 10 years. The rubble from the colossal impact of the two bodies eventually settled down to become just two bodies. The giant impact scenario nicely explains the Moon's lack of a large metallic core, and the presence of similar oxygen isotopes in both Earth and Moon rocks adds weight to the theory. The theory also explains how the Earth captured such a big Moon; it simply got in its way. But if the reader wants to imagine a cosmic creator who decided to conjure up the perfect Earth–Moon system for eclipse chasers, then that is fine by me.

Enjoy Eclipses While They Last

The Moon was originally much closer to the Earth and, even now, it is moving further out. But don't panic, it is only drifting away at a rate of 38 mm per year. How do we know this so accurately? Well, the laser reflectors left on the lunar surface by the Apollo astronauts enable laser beams to be bounced off the Moon and timed with picosecond precision. Obviously, the Moon used to look much larger in our night-time sky and, in the distant future, it will be too small to totally eclipse the Sun. However, this sad day will not occur for 750 million years. Opinions differ on just how close the Moon used to be to the Earth but it does not take a genius to work out that if it has been drifting away from us at 38 mm per year for more than 4 billion years it must have looked twice as big in our skies when the Earth was young and uninhabited. However, astronomers and geologists differ on this point. There simply are not enough solid facts and the Moon may once have been far, far closer to the Earth. Conversely, study of coral growth rings over the last 600 million years suggests that the average drift rate of the Moon, away from the Earth has been nearer 20 mm per year in that period. If this is true it implies that the current continental mass distribution on Earth is much more effective in tidally pushing the Moon away. As the Moon slowly drifts further away from us the Earth's rotation slows down too. It may once have rotated four times faster than today, i.e. the day would have been 6 h long. In the distant future, it is likely that the Earth–Moon rotation will become locked like that of Pluto and Charon. Specifically, the Moon will always sit above the same longitude on the Earth and both will slowly rotate in 7 weeks. However, by this time the Sun will have expanded to become a red giant and so life will have been extinguished.

So the period in which living beings can enjoy the spectacle of an Earth and Moon of the same size in the sky is only a transient one. To me, the fact that we sentient humans can enjoy such an amazing spectacle on what must already be a very rare planet indeed is quite remarkable. How many planets are there with intelligent beings that can also witness such a spectacle? Maybe this is the only one in our galaxy? In addition, and something that hits me every time I see a total solar eclipse, many astronomers now think that the presence of our Moon was crucial to us evolving in the first place. So the very body that may be responsible for this unique and beautiful Earth developing its wealth of biological life-forms puts on an additional show for us by being the same size as the Sun in the sky. Now, I am not a religious man, but that thought does send shivers down my spine. How can the Moon be related to the development of life? Well, the Moon creates substantial tides, far more substantial than if the Sun were the only gravitational tide-raising body. Richard Lathe, a molecular biologist at Pieta Research in Edinburgh, UK, has suggested that the massive (and more frequent) tides raised by the much nearer primordial Moon may have been crucial in the formation of our DNA due to the large salinity level changes produced. Although this is not universally accepted, it is an undeniable fact that the significant tidal effect of the Moon on our oceans has affected the development of all life on Earth. Perhaps, just as crucial is the fact that having such a large nearby Moon orbiting our planet has stabilised the Earth's axis. Without this stability in early geological eras the changes in climate would have been far more dramatic, maybe too dramatic for evolution to cope with.

In passing, it is worth explaining how the Moon's tidal forces are responsible for the Moon moving further away. The tidal bulge raised on the faster rotating Earth, by the Moon, has a subtle tendency to try to accelerate the Moon in its orbit. This accelerative force imperceptibly (before laser rangefinders at least) tends to push the Moon further out into a slower orbit. The Moon's rotation is, of course, effectively locked already by the larger Earth: We always see the same face of the Moon, even if the tiny librations (described later) allow us to peep around the limb now and again.

Defining the Lunar Orbit

The orbit of the Moon around the Earth is highly complex, even if the precise algorithm (devised by various mathematical geniuses over centuries) running on a modern PC can make its motion seem deceptively simple to calculate. During the late seventeenth century, Isaac Newton himself grappled with the problem of the complexities of the lunar orbit and for the substantial total solar eclipse of 3 May 1715 (a 304 km wide track across England) he prepared broadsheet diagrams for the general public. In many ways eclipse chasing started in 1715 as it was the first total solar eclipse for which the British public were well prepared. Eclipses had been predicted without recourse to advanced mathematics since the Babylonian era (640–538 BC) simply by noting that they tended to recur with a period of 6,585.3 days, or 18 years, 11 days and 8 h. So as long as you had a good calendar watcher you were in business almost. The problem is that while this works fine for lunar eclipses, visible from a whole hemisphere of the Earth without any modern solar filters, TSEs occur over a narrow ground track, and that little 8-h add-on shifts the track 8 h in longitude on the Earth's surface, well away from Babylonian observers. It is only in the last few hundred years that truly precise eclipse track prediction has become possible.

Despite Newton's sterling efforts (and great success) in accurately predicting solar eclipses the modern prediction method is based largely on the work of a later genius, the Prussian Friedrich Bessel. In 1824, he looked at the mathematical problem of eclipse prediction from a completely different viewpoint, namely by imagining a flat card passing through the Earth's centre perpendicular to the axis of the lunar shadow striking the Earth. Using this method a mere eight coordinate parameters are needed to define the shadow falling on the Earth's surface. Bessel's new approach made predicting an eclipse ground track much simpler and, even today, planetarium software running on your PC will calculate eclipses using Bessel's technique, albeit with far more accurate orbital elements. Other noted mathematicians, Newcomb, Brown, Eckert, Jones and Clark refined the solar and lunar ephemerides to their current state between 1895 and 1954.

If we simplify the Moon's motion to its most basic elements then we arrive at the following facts:

> The Moon orbits the Earth and returns to the same angular elongation from the Sun in 29 days, 12 h, 44 min and 2.9 s. This is also called the synodic period and, in reality, is an average figure. If you look at the precise time between new Moon and the next new Moon (or full Moon and the next full Moon) it will never be exactly 29.53 days because the lunar orbit is an ellipse and not a circle. However, without going into great detail, every lunar month is roughly 29.5 days long. There is another month too, namely the sidereal month. This is the period the Moon takes to return to the same position with respect to the stars and is 27.322 days long. The two day quicker sidereal month is caused by the fact that the entire Earth–Moon system orbits the Sun in the same anticlockwise (seen from above) direction. In 27.3 days the

Earth–Moon system has itself proceeded along 1/13 of its 365.25 day long orbit, equivalent to that missing 2.2 days between the synodic and sidereal month. Incidentally, to say that the Moon orbits the Earth is not quite correct. They both rotate about the common centre of gravity, i.e. a point just inside the Earth's globe, but not at the Earth's core.

Elliptical Orbits

Not only is the Moon's orbit around the Earth elliptical, the Earth's orbit around the Sun is elliptical too and this is crucial in determining the precise size of Moon and Sun as seen from the Earth. Obviously if the Sun is larger than the Moon a total eclipse cannot take place. In this case a blinding ring of light will surround the Moon and the eclipse will be of the "annular" type: interesting, but not the spectacle offered by totality and the sight of the corona (see Fig. 1.2). The Earth is closest to the Sun around 3 or 4 January each year, which may surprise a few high-latitude northern hemisphere dwellers, who are frozen solid at that time of winter. But of course, it is the Sun's height in the sky that determines the seasons and not its proximity. Six months later, around 4 or 5 July , the Earth is furthest from the Sun. Specifically, we are 147.1 million km from the Sun at our closest and 152.1 million km from the Sun at our furthest. Thus the range of our solar distance is 149.6 million km ± 1.7% (149,597,893 km ± 1.7% to be precise). In terms of the size of the solar disc this can

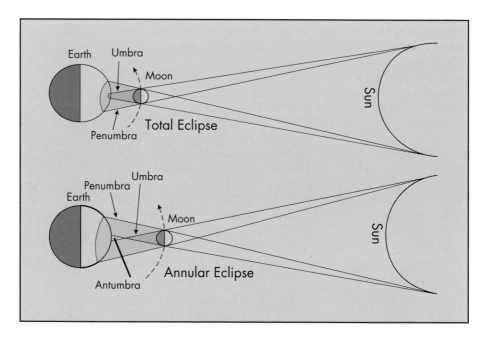

Fig. 1.2. The geometry of a total solar eclipse (*upper*) and an annular solar eclipse (*lower*). From within the penumbra an observer will see the lunar disc obscure part of the solar disc. From within the umbra the whole of the solar disc will be covered. In the annular case the observer would need to be in a spacecraft well above the Earth such that the Moon would appear big enough to cover the Sun. As that is not an option, in practice the observer centrally placed under the shadow will see a ring of light around the Moon. In this case the observer is inside the antumbra.

vary between 32′32″ and 31′28″ or 32′00″ ± 1.7%. This modest variation is considerably less than the percentage that the Moon can change in size (see Fig. 1.3), because its orbit is much more elliptical. Specifically, our distance from the Moon (measured from the Earth's centre) can range between 406,697 and 356,410 km or 381,554 km ± 7%. In fact, the range is even more extreme than this, and in totality's favour. Because the Moon is relatively close to the Earth there is an appreciable advantage in being on the Earth's surface, i.e. up to 6,380 km nearer to it if it is directly overhead. This can make the Moon 1.8% bigger than if the observer were at the Earth's core, or, more realistically, observing the eclipse near sunrise, sunset or with the Sun and Moon low on the horizon. This uplifting effect of the Earth's radius is of critical importance in the so-called hybrid eclipses, where the ends of the track show an annular eclipse, but observers in the middle enjoy totality. The extreme geocentric angular size of the Moon varies between 29′23″ and 33′32″ with an average value of 31′27″. However, if the Moon is overhead, and the observer is at the equator at noon, an absolute maximum angular size of 34′9″ is possible. The Moon's equatorial diameter is 3,476 km and the polar diameter is 3,470 km. A quick glance at the Moon's range of apparent sizes in our sky compared to those for the Sun (see Fig. 1.4) reveals that the Moon can only totally eclipse the Sun when it is closer to us than average; put another way, the Moon needs to be nearer perigee than apogee. In July, when the Sun appears at its smallest, the Moon can almost cover the Sun's disc when the Moon is at its average diameter. However, in January the Moon needs to be significantly nearer to perigee to create a total solar eclipse. Not surprisingly this fact means that annular solar eclipses are slightly more frequent than total solar eclipses. Even when Moon and Sun meet, the Moon is, on average, just a bit smaller than the Sun. Dividing the equatorial diameters tells us the Sun is 400 times bigger than the Moon (1,391,980 km/3,476 km). Dividing the average geocentric distances tells us the Sun is on average about 390 times further away (149,597,893 km/381,554 km). Over thousands of years annular solar eclipses will account for about 33% of all solar eclipses, with total solar eclipses accounting for 27%. Hybrid eclipses (on the borderline between annular and total) account for 5%. The remaining 35% are solar

Fig. 1.3. The relative sizes of the furthest possible Moon, the average Sun and the closest possible Moon.

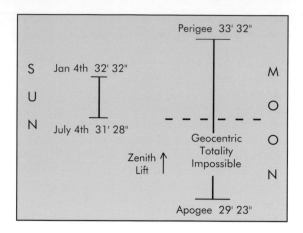

Fig. 1.4. The extreme angular sizes of the Sun and Moon as seen from the Earth. The vertical lines in the diagram demonstrate the range variation in angular diameter. Note how much greater the range in lunar diameter can be. Also note that even when the Sun is at its smallest apparent size, around 4 July, the Moon still has to be above the dotted line (average distance) to create a total eclipse, rather than an annular one. The zenith lift arrow indicates the maximum amount by which the Earth's radius would lift the observer up, and therefore enlarge the Moon, when the eclipse occurs at the zenith.

eclipses that are partials from wherever you live on the Earth. (Of course, all solar eclipses are partials from your perspective unless you are on that critical narrow track.) When the Moon is at perigee, the observer on the equator, and the Sun at aphelion, i.e. in June or July, there is the potential for a truly awesome total solar eclipse. Under these circumstances a totality of 7 min 31 s is theoretically possible. However, do not get too excited. The longest totality in living memory was in 1955 and lasted 7 min 8 s; the longest totality covered in this book is in 2009 and lasts 6 min 39 s. No 7 min totalities will occur in the twenty-first century, but there is a stonker on 16 July 2186, of 7 min 29 s. Wow! Maybe medical advances in the twenty-first and twenty-second centuries will enable some readers of this book to witness that eclipse. An absurd idea? Maybe not . . .

The Five Degree Tilt

So, with the Moon going round the Earth every 29.5 days you would expect an annular, total or hybrid solar eclipse once a month on the Earth's surface, right? If only that were the case, although it would then be very expensive being an eclipse chaser! Unfortunately there is a huge fly in the ointment: the Moon's shadow normally misses the Earth's surface (see Fig. 1.5). If the Earth had a diameter of 90,000 km, rather than its actual (equatorial) diameter of 12,756 km we would grab all of the total, annular or hybrid solar eclipses as the Moon's shadow would always cross the Earth's surface at new Moon. Why does the Moon's shadow usually miss the Earth? Well, the plane in which the Moon orbits the Earth is tilted with respect to the ecliptic (Earth–Sun plane) by just over 5° (5°9′ on average, to be precise, or between 4°59′ and 5°18′). If the orbital plane tilt were just less than 1°, we would

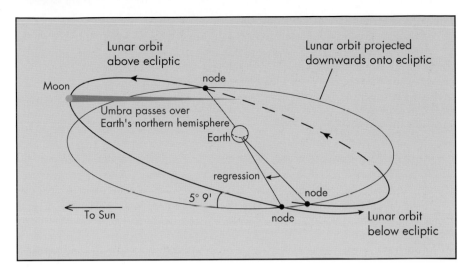

Fig. 1.5. The plane of the lunar orbit is tilted by just over 5° with respect to the ecliptic, i.e. the Earth's orbit around the Sun. In addition, the line of the nodes (the plane intersection points where eclipses occur as Earth, Moon and Sun are in line) regresses clockwise by 19° per year. This regression gradually shifts the two windows per year when eclipses can occur, earlier, by about 19 days per year. As shown in the diagram, when the nodes are at right angles to the Sun–Earth line, the Moon's umbra will sail way above (or below) the Earth at New Moon.

see, from somewhere on Earth, a total, hybrid or annular eclipse at every new Moon, i.e. 12.38 per year and 1,238 per century. However, as things stand Earth dwellers only experienced 150 totals, annulars and hybrids in the twentieth century (71 totals, 73 annulars and 6 hybrids) so in 88% of cases the umbral shadow missed the Earth. To be strictly accurate it is worth adding that there were actually 228 solar eclipses in the twentieth century because there were an additional 78 partial eclipses where the Moon's penumbral shadow touched the Earth, but a total, annular or hybrid could only be seen from a spaceship near the Earth. The figures for the twenty-first century, as calculated by NASA's Fred Espenak, are very similar. In this century there will be 224 solar eclipses, divided into 77 partials, 68 totals, 72 annulars and 7 hybrids. So, if we add the hybrids to the totals we arrive at a figure of 75 potential total solar eclipses, which a long-lived and ultra-keen eclipse chaser could see in the twenty-first century or one every 16 months on average. But is there a good way of mentally visualising how all these eclipse tracks fall across the face of the Earth? Yes, there is, but we need to know a bit more about orbits, nodes and the Saros period.

Orbital Nodes

As we have seen, the inclination of the Moon's orbital plane is, arguably, the major complicating factor in solar eclipse prediction. More often than not the Moon rides above or below the ecliptic and, at new Moon, the shadow falls above the north pole or below the south pole. The optimum situation is when the Moon is situated at, or near, one of the intersection nodes at new Moon, i.e. the point at which ecliptic plane and lunar orbital plane intersect. On the face of it one might expect this to occur twice a year and, in fact, this is not far from the truth. However, the situation is complicated by the fact that the lunar orbital plane itself swivels around in space at a rate of 19° every year, so that node crossings (Sun, Earth and Moon all in line) occur rather more frequently than every six months. In fact, they occur every 173.3 days. The time between the Sun crossing the same node twice is also shorter than a year; it is 346.6 days in fact (the *eclipse year*). Here are a couple of examples to illustrate these two periods. On 3 October 2005 an annular eclipse was seen across Spain, Portugal and the Mediterranean. One-hundred and seventy-seven days later, on 29 March 2006, a total solar eclipse was seen across Africa, the Mediterranean Turkey and Russia/Kazakhstan. The nodes are more than wide enough that the small four day difference (177 − 173.3) did not prevent two consecutive eclipses from occurring. Why 177 days? Well, it is simply six new Moons, i.e. 6 × 29.5 days. Here is another example: On 3 November 1994, a total solar eclipse was seen across Chile, Peru and the Atlantic Ocean. Three-hundred and fifty-five days later, on 24 October 1995, the next total solar eclipse was seen in India and across southeast Asia/Malaysia. In this case 355 days is 12 new Moons (12 × 29.5 almost). Again, the difference between 355 and 346.6 days is not long enough to cause the nodal alignment to fail. In fact, perhaps surprisingly, the nodes are wide enough that two *partial* solar eclipses can be seen at consecutive new Moons. Although this is very rare it occurred quite recently. On 1 July 2000, a low-altitude partial solar eclipse was just visible in the southern Pacific ocean, not far north of the Antarctic (which was in its long winter night, i.e. the Sun was permanently just below the horizon). Twenty-nine days and 7 h later, on 31 July 2000, another partial solar eclipse occurred, this time in the far northern part of the globe, including the Arctic region. The opposite order of events will occur on 1 June and 1 July 2011, i.e. a northern hemisphere partial, followed by an Antarctic-grazing partial (see Fig. 1.6). In fact, if the Moon's orbit was a solid pipe with the same diameter as the Moon it would take between 30 and 37 days (depending on the lunar and solar distances) for the shadow of the orbit to cross the Earth's surface around the time of nodal crossing. Another way of imagining this is that the Sun has to be within a circle 30–37° in diameter centred on each node (degrees and days are roughly equivalent here as the Sun travels 360° in 365.25 days). In the course of any one calendar year this nodal alignment situation decrees that there will be a minimum of two and a maximum of five solar eclipses.

Because the "same-node" crossing interval of 346.6 days is not an exact multiple of the new Moon interval of 29.5 days (It would be a miracle if it was!). This repetition interval between total, annular and hybrid solar eclipses of 12 lunar months (which is very obvious to most eclipse chasers) only tends to last for two or three consecutive years. Following this, after 12 more lunar months, you will often get either a partial eclipse or no eclipse at all. Typically, partial eclipses at that first node eventually start to reoccur one lunar month earlier. Figure 1.7 shows the course of events. During the temporary total/annular/hybrid gap at the one node, the next total solar eclipses occur at the other node, almost 18 months later, or, sometimes, almost two and a half years later. For example, there were total solar eclipses on 11 July 1991 and 30 June 1992, at the same node, i.e. a summer node in those years, 12 lunar months apart. Then there was a wait of 29 lunar

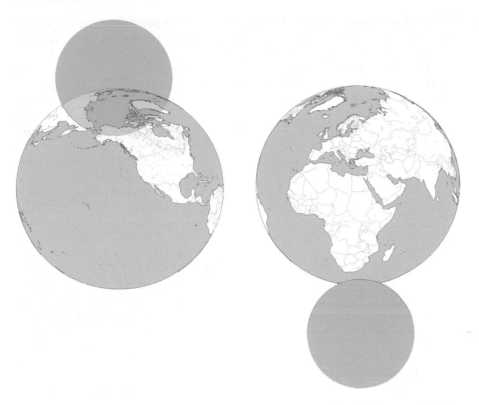

Fig. 1.6. Two partial solar eclipses at successive New Moons are possible, if rather rare. On 1 June and 1 July 2011 partial eclipses will occur at the extreme northern and southern ends of the Earth. Graphic generated using *WinEclipse* by Heinz Scsibrany

months until the next total eclipse at the opposite node, on 3 November 1994. Twelve lunar months later a second total eclipse occurred at the same node, i.e. the aforementioned 24 October 1995 eclipse. But then the next total solar eclipse after that occurred back at the previous node, namely the China/Mongolia eclipse (with comet Hale-Bopp at its peak). Twelve lunar months later, the 26 February 1998 eclipse, viewed largely from the Carribean, occurred at the same node. I would not want the reader to think that my use of the term "node" here implied a fixed date. We have already seen that the two nodes where the lunar orbit and ecliptic plane meet regress back in time by about 19° per year, or, put another way, by 19 days per year. The pattern that the holiday planning eclipse chaser actually tends to notice is the calendar date of successive total solar eclipses drifting back by 11 days (i.e. the difference between 12 lunar months and 365 days) until the third year, when the drift takes the 12th lunar month outside the node and the opposite node, another 6 months later may be more favoured. The string of three total solar eclipses at the same node, in August and July 2008, 2009 and 2010, is actually quite rare. With a partial eclipse the new Moon is drifting out of the nodal zone, but with annulars and hybrids the issue is one of perigee and apogee, i.e. whether the Moon is close to us, or far away. So let us now have a look at this particular aspect.

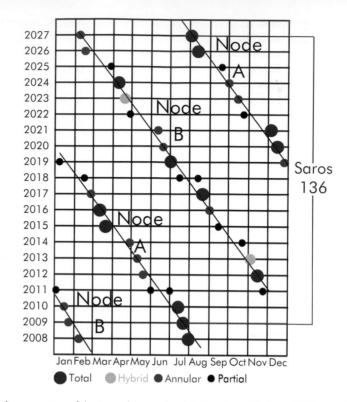

Fig. 1.7. The regression of the two eclipse nodes, labelled A and B, from 2008 to 2027 is evident from the slope of the lines in the diagram. Straight lines represent the centre of the node. Note that as time progresses eclipses first take place at the early side of the nodal window (*left* of the lines), when they are often partial. As they proceed to the centre of the nodal window they are either total, annular or hybrid, i.e. central eclipses. Then, as they move out of the nodal window (*right* of the line) they tend to be partial eclipses again. After this point eclipses typically jump to one lunar month earlier, i.e. they start again on the left-hand side of the nodal centre line. In the case of June/July 2011 (also see Fig. 1.6) and July/August 2018, the earlier lunar month "jump" occurs before the old sequence has finished, hence we enjoy two partial eclipses a month apart. Typically four eclipses occur before the jump to the one month earlier New Moon. The two next very long total eclipses of Saros 136 (numbers 37 and 38) are specifically indicated, i.e. those of 22 July 2009 and 2 August 2027, 18 years and 11 days apart.

The Perigee–Apogee Cycle

If you thought we had dealt with all aspects of the lunar orbit, I am afraid things are even more complicated. As well as the actual position of the Moon with respect to Sun and Earth, the perigee/apogee timings are determined by yet another time period. When you think of the complexities involved here it makes you appreciate the phenomenal mental abilities of Newton when he grappled with predicting the Moon's orbit in the late seventeenth century. Remember, there were no computers or calculators in those days. We have already seen that the lunar orbit is an ellipse with the Earth (or Earth–Moon centre of gravity) at one focus. The shorter radius of the major axis corresponds to perigee and the longer radius to apogee. This ellipse itself slowly rotates, or precesses, forwards (direct motion) completing one rotation in 8.85 years, while the line of nodes rotates backwards (retrograde motion) in 18.61 years. If your brain is not spinning and precessing yet, with all this data, then *you* must be a genius. However, this direct motion of the point of perigee does not affect the time of an eclipse, only its magnitude, i.e. how big the Moon will appear. As mentioned earlier the distance we are from the Sun also plays a part here.

The Saros

At this point, after reading the preceding pages a dozen times, you may think you know all about the underlying periodicity that governs solar eclipses. Certainly, from the eclipse chaser's holiday planning viewpoint the fairly typical "two total solar eclipses at one node, then a gap, then two at the next node" pattern seems very memorable. But this is not the pattern that historians tend to talk about, or eclipse chasers approaching 18 years in the hobby. These people tend to talk in terms of "the Saros cycle". The Saros cycle is of little use to the short-term thinker who has, perhaps, just seen a total solar eclipse and craves knowledge on when the next good one will be. However, it has become part of eclipse language because its 18 year, 11 day (and 8 h) period enabled the really big eclipses to be calculated easily (even in ancient times) and because it takes advantage of two quite remarkable coincidences, at least, remarkable when considered together. (Incidentally, years in this context are calendar years of 365.24 days, so if 5, not 4, leap years occur in the 18 years, the 11 days drops to 10). Now, we have already commented on the utterly remarkable coincidence that the Earth not only has an absurdly large satellite but also appears roughly the same size in the sky as the Sun, about 400 times larger, but about 400 times further away. The other two coincidences are as follows:

Nineteen eclipse years (i.e. 19×346.620077 days = 6,585.78) is only fractionally different to 223 lunar (synodic) months (i.e. $223 \times 29.530589 = 6,585.32 = 18$ years, 11 days and 8 h). In fact, the difference is only 11 h. Now, you could argue that, if you are coincidence hunting and have got to a figure of 19 eclipse years, and over 6,500 days, it is not Earth-shatteringly surprising that you might find a number almost divisible by 29.5. Well, probability theory was never my strong point so I'll leave that debate to any mathematicians reading this chapter. However, it seems quite remarkable to me on its own. But now we come to the second coincidence. It would be very nice, if we have found a pattern after which eclipses (total, hybrid, or annular) repeat if the Sun and Moon were a similar size. That way the eclipses would be very similar ones. In other words, a really stonking 6 or 7 min totality would repeat 18 years, 11 days and 8 h later. Well, astoundingly, that is *exactly* what does happen, because 239 anomalistic months, i.e. the period between successive lunar perigees (27.554550 days) are within 6 h of 223 lunar synodic months ($239 \times 27.554550 = 6,585.54$ days). So, we now know what the Saros is. It is a period of 18 years 11 days and 8 h after which virtually identical eclipses repeat, at a similar time of the year (11 days later) with Sun and Moon, amazingly, almost the same size (see Fig. 1.8). The main difference is that extra 8 h which shifts the track position 8 h westward every 18 years, although, back to the same longitude every 54 years and 34 days, even if the track will have shifted roughly 1,000 km north or south of the original track. Thus a super-dedicated eclipse chaser aged 20 can see a similar eclipse when he or she becomes 74 years old.

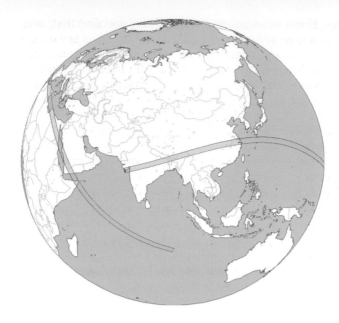

Fig. 1.8. The tracks across the Earth's surface of the two big Saros 136 total solar eclipses of 22 July 2009 (right-hand track) and 2 August 2027 (left hand track). Note how the track shifts by 8 h west after 18 years. Graphic generated using *WinEclipse* by Heinz Scsibrany

After all this mathematics let us look at an example, the best example in fact: the longest eclipses of the late twentieth and early twenty-first century, i.e. ones which occur with the biggest Moon and smallest Sun, therefore peaking in late June and early July (i.e. close to perihelion). Table 1.2 shows the eclipses of Saros cycle number 136. Of course, once we look at the eclipses of this Saros in the distant past and far future, they will be less impressive, as they will move way from solar aphelion, so the disc of the Sun will be larger.

The careful reader will have spotted that, with solar eclipse node alignments occurring every 173.3 days, and with all solar eclipses, partials included, being relatively common events there must be many Saros cycles running simultaneously. In fact there are always between two and five solar eclipses every year, somewhere on the Earth, compared to two eclipses in the same Saros being a huge 18 years and 11 days apart. At the end of the 1900s there were actually 41 Saros cycles still running and almost two-thirds of these were capable of producing total,

Table 1.2. The very long eclipses of Saros 136, namely those from 1919 to 2027

Date	Duration	Eclipse Track
29 May 1919	6 min 51 s	Bolivia/Brazil/Atlantic Ocean/Africa
8 June 1937	7 min 4 s	Pacific Ocean/Peru
20 June 1955	7 min 8 s	Sri Lanka/S. Asia/Philippines/W. Pacific Ocean
30 June 1973	7 min 4 s	Guyana/Atlantic Ocean/Africa/Indian Ocean
11 July 1991	6 min 53.2 s	Hawaii/Baja/Mexico/Central & South America
22 July 2009	6 min 38.9 s	India/China/East China Sea/W. Pacific Ocean
2 August 2027	6 min 22.7 s	Atlantic Ocean/Mediterranean/North Africa/Indian Ocean

The first eclipse shown here, i.e. that of 29 May 1919, which was a member of Saros 136, was a particularly important one. It was used to prove the bending of starlight by the Sun, one of the predictions of Einstein's theory of relativity

annular or hybrid eclipses. If you examine what is happening you find that, over 1,300 years or so, eclipses of a specific Saros cycle gradually slide across and off the Earth's surface, i.e. the Moon's shadow ends up back in space. At the same time new Saros cycles will start as their penumbrae and umbrae hit the Earth's surface for the first time. Saros cycles have a finite duration primarily because of the 11 hour difference between the 6,585.32 day and 6,585.78 day periods. The first eclipses of a Saros cycle start as minor events with the eclipse zone just grazing the north or south polar regions of the Earth (depending on whether the first node is the ascending or descending node). It takes about two centuries (ten or eleven Saros periods) before the central umbra hits the Earth, and the Saros can produce its first total, hybrid or annular eclipses. This happy state of affairs will last for roughly 950 years before the centre of the umbra leaves the Earth's surface at the opposite pole and, after 200 more years, not even partial eclipses will be seen. The whole Saros will have typically lasted some 1,300 years and produced around 75 eclipses including, maybe, fifty which are total, hybrid or annular along a narrow track on the Earth. Needless to say, the best years of any Saros are when the umbra crosses the central equatorial regions of the Earth, maximising the area over which the eclipse can be seen, and with observers raised up just that tiny bit nearer to the Moon. When a Saros provides this around late June and early July, and if the Moon is at perigee, the much sought after 6 and 7 min totalities can be savoured, although, as we have seen, there are no more 7 min totalities in any of our lifetimes, unless any readers of this book make it to 25 June 2150!

Incidentally, if the Earth was as large as 90,000 km in diameter, a figure I mentioned some time back, the individual Saros periods would never end. After many millennia the Moon's shadow would reach a high (or low) point on the massive planet's surface and then move back down (or up). With a planet that big the Moon's orbital inclination is simply not large enough for its shadow to ever miss the Earth's surface at New Moon.

Refining the Prediction

I hope I have explained the mechanisms of solar eclipses reasonably well in the preceding pages. Thinking in three dimensions is never easy, especially when there are a whole host of ellipses, inclined planes and curved surfaces involved! But the countless astronomers and mathematicians who have refined eclipse prediction over the years have to take things very seriously. They want perfection in their calculations and eclipse chasers like to know exactly where an eclipse ground track lies and how long the eclipse will be. This is never more important than when eclipse chasers position themselves at the edge of the track (to refine measurements of the solar diameter) or when an eclipse is a hybrid and the shadow is only a few kilometres wide. To predict solar eclipses to the highest precision extremely accurate astronomical data is required as well as knowledge of how that data is changing with time. As we have already seen, the Moon is drifting away from us at 38 mm per year. But, in addition, the Earth's rotation is microscopically slowing too. Of course, all this accuracy is meaningless unless a universal time standard is employed. The time benchmark of the scientific world is known as U.T.C. or Coordinated Universal Time which, as far as your standard wristwatch is concerned, is equivalent to U.T. (Universal Time) or G.M.T. (Greenwich Mean Time).

In other words 0 h G.M.T. is midnight in London, England, at least in the winter months (British Summer Time, or G.M.T. plus 1 h applies from late March to late October). However, when we require total precision we need to be a lot more careful about using the term U.T. In navigational data and quite a few astronomical almanacs U.T. actually means U.T.1. This is a time measurement linked to the Earth's rotation and the Earth is the only clock that keeps to this time standard. Obviously, in times gone by, the transiting of a star on the meridian, was in keeping with this system. In U.T.1. the Earth is the clock. However, the Earth is not a perfect time keeper. Its rotation is slowing down. The most reliable time keepers used by scientists are atomic clocks. These use the oscillations between an atomic nucleus and its cloud of electrons to count time. The US Naval Observatory (USNO) Atomic Clock consists of fifty caesium clocks and a dozen hydrogen clocks. Frequency data from this ensemble are used to steer the frequency of another clock, called Master Clock 2 until its time equals the average of the ensemble. For those who like definitions, the second itself is currently defined as "the duration of 9,192,631,770 periods of the radiation corresponding to the transition between the two hyperfine levels of the ground state of the caesium-133 atom at a temperature of 0 K". In practice the world's atomic clocks rarely deviate from each other by more than 30 ns during the course of a year whereas the Earth's own U.T.1 clock slows down by about 0.8 s per year. By international agreement, when the difference between U.T.C. and U.T.1. approaches 0.9 s an extra leap second is introduced into the U.T.C. standard to get it back to U.T.1. It may sound odd that the more accurate standard is the one that is adjusted, but if this were not carried out we would have completely abandoned the Earth's rotation as our traditional daily reference system. As it is, we keep the old historical system but with an accurate record of what the real atomic time is. The drift between International Atomic Time (T.A.I.) and U.T.C. is the aforementioned 0.8 s per year. It should be emphasised that this drift is not precisely predictable and it is not solely due to the tidal interaction between Earth and Moon. Short-term fluctuations caused by climatological sea level changes, fluids in the Earth's core, and even (arguably) tidal power stations all play their part. Nevertheless, the Earth continues to slow down and, as yet, leap seconds have never had to be subtracted. I would like to make something clear at this point as this "leap-second" business can cause confusion. The Earth is *not* slowing down/getting out of step by an extra 0.8 s every year. This 0.8 s would be necessary even if the Earth's rotation did not change at all from this point on. The length of an Earth day, as measured back in 1820, and with the second defined in a way equivalent to the current caesium clock standard, is 24 h or 86,400 s. However, the length of the modern day, measured precisely, is 86,400.002 s. It is this extra 2 ms (roughly) that causes the leap seconds to be inserted on a regular basis. Even if the Earth stopped slowing down and settled at its current 86,400.002 average solar day rotation rate, the leap seconds would still be necessary. However, the Earth has always slowed down and may once, billions of years ago, have rotated every 6 h. So, on top of the 2 ms we can probably expect roughly 1.4 ms per solar day per century to be added. So, by 2100 AD we may need a leap second every nine or 10 months or so, instead of every 16 months.

We have digressed quite a bit from eclipses at this point, but how does all this affect precise eclipse calculations? Well, in 1957, the I.A.U. (International Astronomical Union) adopted yet another time system, called Ephemeris Time (E.T.) as the astronomical time standard. In Ephemeris Time the second is defined as 1/31,556,925.9747 of the tropical year 1900. However, by 1984 even this standard succumbed to the superior accuracy of the atomic clocks and a new term, Terrestrial Dynamic Time (T.D.T.) replaced Ephemeris Time, with the new system synchronized to the Ephemeris Time of 1977. Solar Eclipse predictions are now also based on the atomic accuracy T.D.T. system. T.D.T. is equal to International Atomic Time (T.A.I.) plus a fixed offset of 32.184 s. T.D.T is also equal to U.T.C. + 32.184 s + the number of leap seconds since they were first introduced on 1 January 1972. At the time of writing (2006) the 33rd leap second has been introduced so TAI minus UTC is now 33 s and TDT minus UTC is 65 s. Eclipse predictions are nowadays routinely calculated using the atomic clock based T.D.T., but obviously the end result has to be in a format based on U.T.C. as no-one (at least no-one I know, and I know some real nutters!) has a wristwatch that runs 65 s fast. Of course the fact that the slowdown in the Earth's rotation cannot be precisely predicted does mean that tiny errors can creep into the position of an eclipse ground track, at least in longitude. A half second error in time can lead to a quarter of a kilometre of error at the equator, but this is fairly trivial when you consider how jagged the lunar limb is. However, when trying to predict total solar eclipse tracks decades ahead, errors of a few kilometres in the tracks' longitudes will result unless the slowing down of the Earth's rotation is guesstimated accurately.

According to Fred Espenak's 1987 Canon of Solar Eclipses (1986–2035), in practice the uncertainty in the umbra's position using modern computational ephemerides amounted to a kilometre in the north/south direction, 2 km in the east–west direction and just over 2 s of time. That is without any unexpected slowing in the Earth's rotation.

Chapter 2

The Solar Atmosphere

From the viewpoint of the newcomer to total solar eclipse chasing, it might be thought that the Sun simply disappears when covered by the Moon, i.e. everything goes black. However, the real beauty of the sight of an eclipsed Sun is in seeing parts of the solar atmosphere that are normally hidden. Above all, it is the solar corona that is the most awesome. While modern H-Alpha filters (see Chap. 13) can show you solar prominences on any sunny day, albeit in a weird shade of deep red, viewing the corona is still unique to total solar eclipses.

The Photosphere

Let us take a look at what our nearest star actually consists of. The brilliant solar surface which is seen when you project the Sun onto white card, or when you use, e.g. the appropriate Neutral Density 5 (1 part in 10^5 gets through with no colour bias) solar filters, is called the photosphere (a Greek-derived word meaning "sphere from which light comes"). It is tempting to think of the photosphere (Fig. 2.1) as being a solid surface below which the gaseous solar furnace resides. In fact, apart from the central core, we can realistically regard the Sun as being a huge incandescent, unbelievably hot, gas ball, up to the photosphere, at the photosphere and even beyond the photosphere. We simply see the photosphere as if it were a solid surface because it is opaque and we are looking straight through the gas that lies above it. But the Sun does not have a solid surface. It is just one big atmosphere and the parts that stretch way outside the surface-like photosphere are what we see during a total solar eclipse. If only the photosphere was transparent and astronomers could glimpse the inner machinery of the Sun! In fact, that blindingly bright impenetrable top surface is only about 400 km thick. On a globe almost 1.4 million km in diameter the photosphere is, relatively speaking, thinner than the skin of an apple. Nevertheless, it is what Earth based astronomers see and it contains the famous "rice-grain" granulation visible even in good amateur images. This granulation is the result of supersonic gas convection "bubbles", roughly 1,500 km across, bursting and cooling at the visible surface. The lines between the granules are where the cooling material sinks back down. The photosphere is not a surface in the same sense as the Earth's surface. It is merely an opaque cap on the top of the convective zone beneath. The sunspots (Fig. 2.2), so well known to all solar observers, are, perhaps surprisingly, regions on the solar surface where intense magnetic fields have reduced the energy arriving from the convective layer beneath. Sunspots are therefore considerably cooler and less

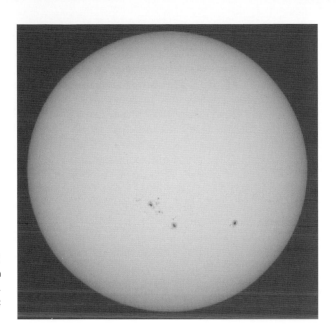

Fig. 2.1. The Sun on 21 June 2004. Vixen 80 mm f/8 Apo + Fuji S2 Pro DSLR. Baader solar filter. Image: Damian Peach

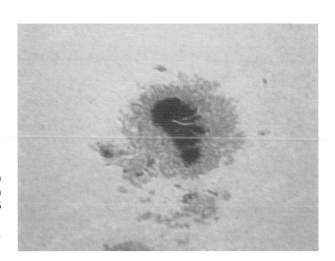

Fig. 2.2. A large sunspot on 17 July 2004. Vixen 80 mm f/8 Apo at f/45 + Atik 1HS webcam. Baader solar filter Processed with Registax. Image: Damian Peach

bright than the rest of the visible solar surface, and their darkest parts represent the coolest and most magnetically active areas. Sunspots and the effects they generate are just like bar magnets and iron filings in those early physics lessons at school. They group together in twos, one with a positive polarity and the other with a negative polarity, usually in an east–west pairing. Magnetic arches join the two sunspots and the solar equivalent of the bar magnet's iron filings manifest themselves as so-called prominences when the Sun's rotation takes them to the solar limb. When viewed from above, i.e. in the centre of the solar disc, they appear as dark, snake-like filaments at the hydrogen-alpha wavelength.

Prominences

Prominences at the solar limb, especially when the Sun is very active, can be truly spectacular (see Chap. 11 for eclipse images of prominences). The largest prominence ever recorded was on 4 June 1946. The photographs of this event, recorded at the Climax Observatory in Colorado, amazed the scientific world. That prominence was known as the Grandaddy prominence and it extended 200,000 km above the solar surface and stretched across more than half a million kilometres of the solar limb. Such prominences are extremely rare and so the chances of a really large one being visible during the few minutes of totality at a total solar eclipse are very slim. Nevertheless, at the total solar eclipse of 29 May 1919, famous for its role in proving Einstein's theory that gravity bends light, a superb prominence was photographed by the British Eclipse Expedition led by Arthur Eddington at Principe Island (in the Gulf of Guinea, off the west coast of Africa). While not as spectacular as the awesome Grandaddy prominence of 1946 it was an awesome prominence to be seen at a total solar eclipse and nothing larger has been seen at an eclipse since. The 1919 prominence towered some 100,000 km above the solar surface. What a sight that must have been! Large solar prominences tend to occur more when the Sun is very active and Sunspot numbers are high. As we shall see shortly, the Sun has a 11-year period of activity.

Vassenius, the Swedish observer, may have been the first to describe the appearance of prominences in any detail, during a total solar eclipse in 1733. However, the observer, Captain Stannyan, observing from Berne in Switzerland, may have seen prominences 27 years earlier. When John Flamsteed presented a report of the 1706 total solar eclipse to the Royal Society, he mentioned Captain Stannyan's observations, in particular, his comment: "The Sun getting out of his eclipse was preceded by a blood-red streak of light from its left limb, which continued not longer than six or seven seconds of time." Flamsteed's comments on this observation were: "The captain is the first man I ever heard of that took notice of a red streak of light preceding the emersion of the Sun's body from a total eclipse. And I take notice of it to you because it infers that the Moon has an atmosphere." However, speaking as someone who has witnessed five totalities it sounds to me as if Stannyan may well have been describing the Sun's chromosphere (described shortly), as that also could be described as a red streak emerging for a few seconds before the solar photosphere. Vassenius, possibly influenced by the ideas of the time (and maybe even Flamsteed's comments of 27 years earlier) thought the prominences were a lunar phenomenon too. After the August 1868 eclipse, Lockyer (England) and Janssen (France) worked out on how to observe them spectroscopically, even when an eclipse was not in progress.

Although the sunspots, visible in white light, are cooler than the surrounding photosphere they are still very hot and as bright as an arc lamp, in real terms. Their darkness is just a relative effect. The temperature of the photosphere is typically about 5,800 K compared to around 3,500 K for a sunspot. The region surrounding a sunspot can appear bright, especially when a spot is near the solar limb, as seen in white light. These regions are called faculae. In narrow-band H-Alpha filters large regions of the solar disc, surrounding sunspots, can appear white; these features are called plages and are in the thin layer just above the photosphere called the chromosphere (shown in Fig. 2.3).

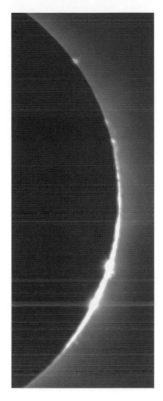

Fig. 2.3. The solar chromosphere is briefly visible at second and third contacts during total solar eclipses. This image (1/500 at ISO 400) was taken at third contact on 29 March 2006 using a Canon 300D digital SLR and Celestron C90 (90 mm aperture f/11). Image: M. Mobberley

The Chromosphere and Spectral Lines

Features in the chromosphere can be observed at deep red hydrogen-alpha wavelengths (6,563 Å) or in the violet wavelengths of singly ionised calcium (3,933 and 3,967 Å). In passing I would just like to digress and clear up a point here that confuses many non-physicists. You may well ask, "What the hell has calcium got to do with the Sun?" Looking at the Sun at the wavelength of hydrogen-alpha seems quite reasonable; after all, the Sun consists mainly of hydrogen. But I think most people will automatically associate calcium with teeth, bones and milk. So what on Earth is that element's connection with the Sun! Although most amateur astronomers have some appreciation of the fact that elements can be identified as vertical lines in a star's spectrum, a brief explanation of the origin of spectral lines may not be inappropriate at this point. A bright hot star emits radiation across the electromagnetic spectrum, and when an optical prism, a diffraction grating or even raindrops are used to disperse the light into its component colours, we see the standard rainbow colours from red to violet. However, when the light in question has passed through a gas on its way to us, photons may be absorbed by electrons orbiting the atomic nuclei in the gas and the electrons will then jump up a discrete orbit level as they absorb photons. This absorption leads to dark lines appearing at discrete wavelengths in the spectrum. If I want to be totally accurate here I would have to admit that when photons are absorbed and an electron moves to a new energy level, the electron eventually returns to its original level, re-emitting a photon. So you might think the effect would cancel out. In fact, because the photons can be re-emitted in any direction but were absorbed while heading straight for us, there is a net dimming of the overall light, i.e. the dark absorption line.

The opposite effect occurs when the gas is being excited by some energy input. In this case the electrons may jump down a discrete orbit level as they emit photons. This emission leads to bright lines appearing in the spectrum. The well-known Balmer series of hydrogen atom orbit transitions give rise to lines in the visible part of the spectrum and correspond to electron transitions between the second orbit level and higher orbit levels. It was only with the development of quantum physics and probabilities that the discrete allowable orbits were fully understood.

Using narrow-band filters to study the Sun at specific wavelengths shares the above principles. At the frequencies corresponding to these orbit transitions a great detail of contrast can be seen and, depending on the frequency, a different layer of the star's atmosphere can be studied. Of course, if an orbit transition results in absorption or emission outside the waveband that our eyes are sensitive to then we cannot see it. The same applies to CCD detectors although their spectral range is slightly wider than that of the eye. It just turns out that the corresponding hydrogen-alpha and calcium wavelengths are the two frequencies for our Sun that are within that region. However, you may still have some suspicions about calcium, despite what I have just said. After all, it surely cannot be a major constituent of the solar atmosphere. Dead right! In fact, there is estimated to be almost half a million times more hydrogen than calcium in the Sun. However, the strength of absorption by a certain element is dependent not only on the amount of a particular element but also on its absorption efficiency. Hydrogen has low-absorption efficiency (dependent on electron availability and the likelihood of absorption when a photon whizzes by), whereas calcium has a very high absorption efficiency and the line(s) produced are still within the human visual range, although well into the violet region.

Anyway, just to repeat my sentence of a while back, features in the chromosphere can be observed at deep-red hydrogen-alpha wavelengths (6,563 Å) or in the violet wavelengths of singly ionised calcium (3,933 and 3,967 Å). Today, narrow-band filters revealing only these wavelengths are well within the financial reach of the keen amateur.

The chromosphere (from the Greek *chromos* meaning colour) lies immediately above the Sun's visible surface (the photosphere). Even with appropriate filters it is not normally visible to observers but, at total solar eclipses it is visible for a matter of seconds just after the last brilliant point of the photosphere disappears (second contact) and just before it reappears (i.e. just before the so-called "diamond ring" effect at third contact). Visually it has a vivid pinky red colour and appears as a dramatic curved line just above the lunar limb at the point where the photosphere will emerge. The chromosphere has a depth of about 2,500 km so it is considerably thicker than the photosphere although still only a skin on the apple. Ultra-high resolution images at hydrogen-alpha wavelengths show that the outer edge of the chromosphere, when looked at from a shallow angle, i.e. the foreshortened limb view, resembles a spiky mountain range. Indeed, the spiked extensions are actually called spicules. These spicules are short-lived jets of gas (they only live for a matter of 5 or 10 min), travelling out from the main body of the chromosphere. They can shoot up to more than 10 km in height (so the mountain peak analogy is valid) and then fall back to the average chromosphere surface, just a few minutes later. In the best images they resemble mountains of iron filings on a sheet of paper with a bar magnet underneath. Indeed, solar physicists think that the spicules may be involved in conveying magnetic fields out from the Sun to heat the next region of the solar atmosphere, the corona.

The Solar Corona

Unlike the photosphere and the chromosphere the solar corona (Fig. 2.4) is not limited to a thin skin-like surface on top of the seething cauldron of energy, we call the Sun. The corona extends to millions of kilometres into space and is the most awesome sight to behold during the totality phase of a solar eclipse. I have heard the black disc, nestling within the electric blue corona, described as looking like the eye of God. (In case you are getting confused, the different components of the Sun are labelled in Fig. 2.5.) While the Sun does not have a solid surface, and even the white-light visible photosphere is really gaseous, the corona actually does look gaseous when viewed at totality. Most first-time eclipse watchers, technically known as "eclipse virgins", are amazed at the brightness of the inner corona. My first

Fig. 2.4. This is a composite of the solar corona and disc at the time of the 29 March 2006 total solar eclipse. It is probably very similar to the visible appearance which would be seen if it was possible to completely switch off the dazzling solar photospheric radiation. The image was created from a series of fine eclipse images by Peter Aniol, expertly processed by Miloslav Drückmuller. The black disc of the Moon has been replaced by a SOHO EIT 17.1 nm image (Fe IX/X) taken 20 min after the eclipse images were taken in Libya. The corona images from Libya were taken with a 100 mm aperture f/8.2 Takahashi ED refractor with a 2x converter, giving f/16.4. A Canon EOS 5D II digital SLR at 100 ISO was used for the 60 exposures which ranged from 1/1000 to 4 s © 2006 Miloslav Druckmüller, Peter Aniol, ESA/NASA.

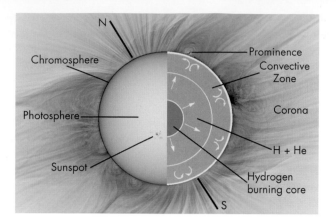

Fig. 2.5. The Sun and its internal and external components, a labelled composite of Figs. 2.1 and 2.3. Nuclear reactions on an enormous scale take place in the centre of the Sun, the core of which is thought to be at a temperature of some 15 million Kelvin. Energy flows out from the centre and supports the enormous mass of the outer layers. Just below the visible surface lies the convective zone. The blindingly brilliant yellow surface which we see (when filtered by 100,000 times with safe solar filters) is called the photosphere and is often peppered with sunspots, especially near solar maximum. Above the photosphere is the thin chromosphere seen at total solar eclipses at second and third contacts (see Fig. 2.4). The photosphere has a temperature of roughly 5,500 K. Prominences arching above the limb of the Sun can be seen at total eclipses or when using suitable hydrogen-alpha filters. Stretching out from the Sun and only visible at total solar eclipses is the tenuous solar corona which has the mind-boggling temperature of 1 million Kelvin.

glimpse of it was for a split second at the 1991 total solar eclipse in Hawaii. Essentially, I was clouded out, *but* as the cloud thinned for a split second there was the merest fleeting glimpse of a dark circle surrounded by a white ring. That was all I saw, but my immediate (and absurd) instinct was "they've got it wrong, this is an annular eclipse!" Admittedly, I was viewing the eclipse through cloud, but I was confused by the bright ring around the Moon. Even the inner corona only glows at a level of a million-fold below that of the solar surface but this is bright enough for it to dazzle like the full Moon. The corona is bright because it, or rather the free electrons within it, scatters the intense light of the Sun. Obviously, its intensity fades as one moves further from the blindingly bright photosphere, but even the human eye can trace the solar corona out to a distance of four or five solar radii beyond the lunar limb. In fact, the most tenuous extremes of the solar corona spread throughout the solar system. Astronomers call the point where our Sun's solar wind hits the cooler background of the interstellar medium, the heliopause and this is currently estimated to lie about 23 billion km from the Sun. Of course, the corona we see at a total solar eclipse is only a few million kilometres from the Sun. Its intricate gossamer structure is, once again, all tied up with the intense solar magnetic fields. It may come as something of a surprise to learn that the corona has a very high temperature, much higher in fact than the photosphere and chromosphere. This seems at odds with common sense. How can a tenuous region of gas so far from the solar centre have a higher temperature than the actual solar surface? It is important to remember here that there is a huge difference between the definition of temperature and heat capacity in the physical world. The temperature of a gas is determined by its average kinetic energy; in other words, the speed that the particles are moving is critical. Indeed, the kinetic energy is proportional to the speed of the particles squared. Gas at a low temperature will not have the energy to stray far from the immense gravitation at the surface of the Sun. So the very fact that we can trace the corona out to great distances tells us that the particles are moving at huge speeds and therefore are, by definition, at a very high temperature. In fact, the temperature of the corona is a staggering one million degrees, more than two orders of magnitude than that of the visible photosphere. In passing, it is worth mentioning that physicists and solar astronomers, even up to the Second World War, were so confused by the spectrum of the solar corona that they thought it might be made of a new element, coronium. It was only when Edlén and Grotrian proved that the coronal emission lines were caused by ordinary elements deprived of a dozen or so electrons per atom that coronium was resigned to history. Why were they deprived of so many electrons? Well, simply because at a million degrees, many electrons escape from their atoms and electrons in the corona are moving so quickly that they will dislodge other electrons from their atoms when they collide. The corona is at such a high temperature that it actually emits most of its energy as X-rays, i.e. high-energy photons. This is why instruments onboard the highly successful SOHO spacecraft (http://sohowww.nascom.nasa.gov/) were designed to observe the Sun in X-rays. But even though the temperature of the corona has been verified by various scientific methods, the reason it is at a million degrees, more than 100-fold higher temperature than the photosphere and chromosphere beneath it, is still puzzling solar physicists. After all, the heat is generated inside the Sun. So, instinctively, one might quite reasonably expect the solar atmosphere to get cooler as one moves further out. The best theory is that intense and changing magnetic fields are responsible for the coronal heating. However, pinning down the precise mechanisms responsible is very difficult. One possibility is that the turbulent magnetic fields rising from beneath the photosphere might generate electric currents that flow through the corona like a battery connected to a thin piece of wire. But one thing is certain: The temperature of the corona is not due to heat rising from the solar surface. The corona is so tenuous that the vast majority of the Sun's light and heat just pass straight through it. So, at a total solar eclipse, what we are seeing when we look at the subtle and complex patterns in the electric blue corona is a tiny fraction of a percentage of the light from the Sun scattered in our direction by the coronal electrons. The intricate patterns are formed by the intense magnetic fields in the region of the Sun. The corona nearest the lunar limb is dazzlingly bright, simply because of its very close proximity to the intensely bright photosphere, but as we move outwards the corona dims and we can see the delicate structure within it. At around five solar radii from the Sun the features get increasingly subtle as the corona merges into the twilight sky seen at total solar eclipses. In fact, as we have seen, the corona extends much further than this if you have sensitive enough X-ray instrumentation to detect it.

The 11-Year Cycle

Although the shape of the solar corona is different at every eclipse, it does tend to have a different general shape depending on whether the Sun is at the peak of its 11-year activity cycle or at the solar minimum point. Solar maxima, when the Sun is very active and has hods of Sunspots on its face, occurred in 1906/1907, 1917, 1928, 1937, 1947, 1957/1958, 1968/1969, 1979/1980, 1990/1991 and 2000/2001. The period of 11 years is not all that regular though, which, personally, has always seemed a bit worrying to me. I guess my simple mind just expects something as huge and essential to life as the Sun to be a bit more trustworthy. There was a 17-year period between the solar maxima of 1788 and 1805 and just over 7 years between the maxima of 1829/1830 and 1837. However, in the twentieth century the most extreme variation was the 9 years from 1928 to 1937. Solar minima occur, not surprisingly, about midway between the maxima. At the start of a new solar cycle, i.e. just after the minimum, the first sunspots appear at latitudes of roughly between 30° and 45° (both sides of the solar equator). Eventually, spots occur nearer to the equatorial regions and the average sunspot latitude at maximum is about 15°. The spots then die out, before they reach the equator and even as the last spots disappear the higher latitude spots of the old cycle begin. The whole cycle can be illustrated nicely with a so-called "Butterfly Diagram", invented by Maunder in 1904. Of course, during totality, the Sunspots can never be seen. However, during the partial phases it is always nice to have a few big sunspots to measure the Moon's progress across the disc with reference to.

The number of sunspots on the disc is the white light solar observer's indicator of where we are in the solar activity cycle. It is also an indicator of how the magnetic activity is taking place over the solar surface. However, for eclipse chasers the shape of the corona is usually a very reliable indicator of solar activity too. At solar maximum the corona extends all around the solar disc, at every angle, from the sun's poles through the solar equator. But there are usually no ultra-long thin coronal streamers. Conversely, at solar minimum, there tend to be wide, long, and obvious east–west streamers, i.e. streaming out in a long thin plane parallel to the solar equator. The corona is no less spectacular at solar minimum, it just looks as if all the magnetic energy is being funnelled into making the equatorial streamers.

Chapter 3

The Shadow of the Moon

So far, although we have discussed precisely how eclipses occur, we have only looked at the circumstances from a position in space, i.e. by examining the relative positions of Sun, Earth and Moon from afar. We have also discussed the parts of the solar atmosphere that we can see while the Moon is hiding the Sun. But, for me, one of the most breathtaking events at a solar eclipse is what happens to the twilight eclipse sky (see Fig. 3.1) in that last minute before second contact, when it is as if the Gods are turning down a huge light dimmer in the sky. It makes the hairs on the back of your neck stand up. It is the approach of the edge of the shadow of the Moon, speeding towards you at, typically, around 2,000 km/h.

Umbra and Penumbra

Shadows created when an extended source (i.e. not a point source) is obstructed, consist of two parts; an umbra and a penumbra. Within the umbra of the Moon you cannot see the solar surface, i.e. the Sun is totally covered; but within the penumbra you see part of the solar surface, maybe 99% at the early stage or 1% approaching second contact. So for a total solar eclipse you have to be within the umbra, or, put another way, the umbra has to reach the Earth's surface where you are standing. In the best possible total solar eclipse, when the Moon is closest to us (lunar perigee) and the Sun furthest from us (solar aphelion), the lunar umbra is actually long enough to pass 23,500 km past the centre of the Earth's core. In this situation the umbra can have a maximum diameter of 273 km just before it hits the Earth's surface. In the worst possible situation, or, the biggest "ring of fire" annular situation the umbra cone ends 39,400 km before the centre of the Earth, or roughly 33,000 km short of the Earth's surface, if the eclipse zone fell near the equator (see Fig. 3.2). In this latter case, an observer situated on a geostationary satellite directly above the equator might just be able to witness totality. A personal, movable, geostationary spacecraft would be a useful toy to have for changing annular eclipses into total eclipses. The maximum diameter of the antumbral cone, i.e. the zone in which an annular eclipse is seen just before it strikes the ground, is 374 km.

Fig. 3.1. The edge of the umbral shadow is obvious in this picture supplied by Ray Emery, taken during totality, and looking towards the horizon on 29 March 2006 in the Libyan Sahara.

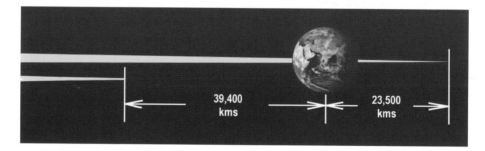

Fig. 3.2. The extreme conditions of the Moon's shadow. With the Moon at apogee (furthest and smallest) and the Sun at perihelion (nearest and biggest) the umbra stops 39,400 km before the centre of the Earth. Thus, this would give the biggest "Ring of Fire" annular eclipse for an observer situated below the umbral tip, i.e. within the antumbra. With the Moon at perigee (nearest and biggest) and the Sun at aphelion (furthest and smallest) the umbra can sail 23,500 km past the centre of the Earth. For an observer on the Earth's surface within the umbra the Moon, at the zenith, could be as much as 8% bigger than the Sun leading to a very long total eclipse.

The Shadow on the Ground

During total solar eclipses all observers on the centreline of the shadow track will experience a totality lasting from zero seconds in duration (for a borderline hybrid eclipse, with an infinitesimally thin track width) to seven and a half minutes duration. Away from the centreline the duration of totality will drop to zero seconds as one approaches the track edges.

The formula for calculating the duration away from the centreline is Duration = Centreline duration $\times \sqrt{[1-(2D/W)^2]}$, where W is the width of the track and D is the distance from the centreline (at right angles to the track). Although most observers will want to be bang on the centreline, sometimes that is not geographically practical and sometimes the coaches get stuck in a traffic jam on eclipse day. So it is useful to be able to calculate the effects of being "off centre". Fortunately, straying a short distance from the centre of the shadow does not affect the duration too much. For example, imagine a 100-km wide track, such that there is 50 km from the track centre to the track edge. At 10 km off the centreline and 40 km from the edge, the duration of totality will be 98% of the centreline value. At 20 km from the centreline you will still see 92% and at 30 km, 80%. Even 40 km off the centreline and 4/5 of the way to the edge the duration is 60%; at 45 km it is 44%. Finally, at the edge, the duration of totality is, of course, zero. As long as you are nearer the centre of the track than the edge you will see 87% or more of the maximum duration.

When you witness a total solar eclipse you must, by definition, be inside the umbra. If the Sun and Moon appear to be almost the same angular size in the sky then that umbra will be vanishingly small. In practice, this means that the track of totality across the Earth must be very narrow. We saw, in Chap. 1, that sometimes a hybrid eclipse occurs where the start and end points of the track feature an annular eclipse and the centre point, raised up by the bulge of the Earth, features a total eclipse. Of course, such borderline total solar eclipses are complicated by the fact that the lunar limb is not a perfect circle some 3,476 km in diameter. In fact, it has a mean equatorial diameter of 3,476 km, a mean polar diameter of 3,470 km, and mountains which can tower many kilometres above the surrounding terrain. When the Sun and Moon are the same diameter in the sky, the umbral track vanishes to zero width. Perhaps the most extreme example of this in modern times was the eclipse of 3 October 1986, over the North Atlantic. The track was a narrow crescent shape, near sunset, from Iceland to Greenland and the duration of totality was predicted as zero seconds. Those eclipse chasers who saw it witnessed a Moon which briefly flashed as a circle of Baily's Beads all around the limb, and then totality (if that is what it was!) was over. There is a page on witnessing this unique experience at Glenn Schneider's website at http://nicmosis.as.arizona.edu:8000/ECLIPSE_WEB/ECLIPSE_86/ECLIPSE_86.html.

I will have more to say about the role of the rugged lunar limb, when Sun and Moon are almost the same size, in Chap. 4.

The Twilight Sky

When you are within the umbra the view you experience varies wildly from eclipse to eclipse. Obviously when the Moon looks much larger than the Sun the shadow will be large and the twilight sky you experience will be a dark one. But the eclipse chaser's sky is never a nighttime sky. As we have just noted, the umbra diameter just before it strikes the Earth's surface is 273 km wide, at best. I have chosen my words carefully here because the umbral shadow rarely hits the Earth's surface directly from above. Unless the Sun is overhead in the observer's sky, the shadow cone will hit the Earth's surface at a glancing angle. Thus if the

Sun and Moon are low down near sunrise/sunset the shadow will have a pronounced east–west elongation on the Earth's surface. If it is low down near the south meridian, e.g. for a noon eclipse viewed from a high northerly latitude, the umbral shadow will be distinctly north–south elongated. Likewise for a low altitude, north-meridian crossing, southern hemisphere eclipse. However, even with an umbra on the ground, stretching hundreds of kilometres, the sky at totality is still far from being a nighttime sky. This often surprises first timers but it should not surprise anyone familiar with even the basics of astronomy. In a typical evening, at the point when the Sun is up to 6° below the horizon (roughly half an hour after sunset), we are at the so-called "civil twilight" and, typically, it is when the streetlights will have started to turn on. Then we enter a period called "evening nautical twilight", which officially ends when the Sun is 12° below the horizon, or about an hour after sunset. Finally, after nautical twilight, we have evening astronomical twilight. This is defined as ending when the Sun is 18° below the horizon, or about 2 h after sunset. The sky is then as dark as it will ever get, even from the countryside. So the Sun has to be way below the horizon before the last vestiges of its light cannot filter through into the upper atmosphere. In a total solar eclipse the Sun is merely covered by a small circle almost the same size and not by the entire body of the Earth. Sunlight is still striking the Earth's atmosphere somewhere within 140 km of the shadow centre as it hits the upper atmosphere, even in the best total solar eclipses. This, and the light from the solar corona, creates an illumination level similar to that at the boundary between civil and nautical twilight, i.e. when the Sun has set and is 6° below the horizon. Of course, normal twilight is different because the brightest glow is just above the point where the sun set (or is about to rise, at dawn). In a total solar eclipse the glow is all around the horizon and will be less intense along the long axis of the shadow. So if the Sun is low in the sky and due south, the elongated north–south ellipse of the shadow will make the east and west horizons appear brighter. Of course, the glowing corona, similar in brightness to the full Moon, slightly modifies the illumination too.

I have heard people say that, during a total solar eclipse, totality was "like suddenly plunging into a tunnel as the edge of the Moon's shadow swept overhead". Personally I have never experienced that. Indeed, neither has anyone I have travelled with. However, when the Sun is at a low altitude and the umbral footprint is very wide, fish-eye lenses have recorded a tunnel-like appearance (see Nigel Evan's Fish-eye shot, Fig. 24 in Chap. 11). The shadow of the Moon is not an abrupt boundary. If the Sun were a point source it would be, but the penumbra of the Moon's shadow is 3,000 km wide either side of the umbral shadow. Even 3,000 km from the zone of totality a partial eclipse can occur where the Moon nicks the edge of the solar disc. Thus, as the Moon's shadow sweeps over you, the illumination level does not go from 100% sunlight to 100% nighttime. At the critical moment the penumbra simply gets darker as it merges into the umbra. Every part of the atmosphere you can see at that time is under the penumbra or umbra and even 300 km from the umbral shadow's edge only a few percent of the normal illumination level is present. So, day does not change to night; twilight changes to dark twilight. Thus, the actual arrival of the shadow edge is quite subtle and blurred. However, in those last 30 or 40 s before totality the drop in illumination level is very spooky, changing by the second, and too rapid for the eye to dark adapt too. You may not be able to see a distinct sharp edge to the shadow but the whole sky fades like a God is turning down a giant dimmer switch in the sky.

Visualizing the Shadow

It can sometimes be quite confusing, when you arrive at an eclipse site, trying to picture exactly how the shadow itself is going to sweep over your location. You may be seeing an eclipse just after sunrise, at midday or just before sunset. But how do you visualise which direction the umbral shadow is coming from? The shadow of the Moon essentially travels across the face of the Earth from west to east, even if this track can be angled and curved substantially by the Earth's axial tilt and high geographic latitudes. This easterly motion is a result of the Moon's mean orbital velocity of 3,682 km/h (61 km/min) in an easterly direction resulting in its shadow travelling at a mean 3,380 km/h with respect to the Earth's core. Of course, the Earth's surface also rotates in the same direction, at a velocity between 1,670 km/h at the equator (28 km/min) and 0 km/h at the precise north or south poles (see Fig. 3.3). However,

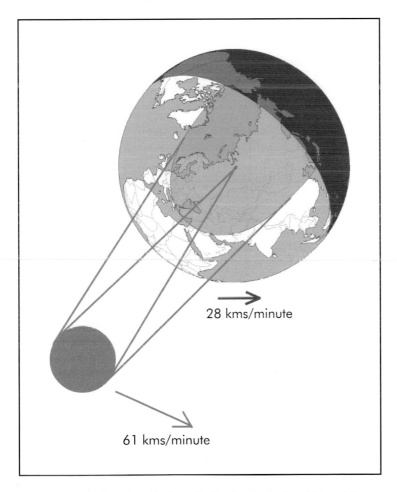

28 kms/minute

61 kms/minute

Fig. 3.3. During a total solar eclipse the lunar shadow hits the Earth's globe and then moves from sunrise to sunset over the surface, i.e. heading, roughly speaking, from west to east. The speed of the shadow across the Earth is largely a product of the Moon's speed in orbit (61 km/min) minus the speed of the Earth's rotation (28 km/min on the Equator). The diagram shows the eclipse of 1 August 2008 just after the track enters Russia. The Earth–Moon distance is *not* to scale. Globe graphic generated using *WinEclipse* by Heinz Scsibrany

even at the equator, the Moon's shadow is moving much faster than the ground, averaging 1,710 km/h, and the eclipse proceeds from where the shadow first hits the Earth's globe, at sunrise, to where it leaves the globe, at sunset. Typically, even at an excellent total solar eclipse, the umbra is only on the Earth's surface for just under 4 h: the time it takes the Moon's shadow to cross a 12,756 km diameter sphere (12,756 km/3,380 km/h). The various north, south, east and west boundaries featured in global eclipse shadow maps are explained in Fig. 3.4. In the middle

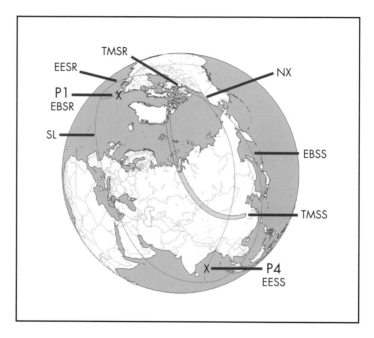

Fig. 3.4. Various curves and lines appear on global eclipse diagrams and they can be rather confusing at first glance. Essentially they delineate the north/south limits of the penumbral shadow (i.e. the partial eclipse zone), the limits of the umbral shadow (i.e. the narrow total eclipse track), and the places at the sunrise and sunset ends of the eclipse where observers just miss out on seeing even the tiniest eclipse or, see the whole eclipse while the Sun is just above the horizon. In the diagram (again for the 1 August 2008 eclipse, where there is no northern penumbral limit on the Earth's surface, as the shadow is so far north) the following abbreviations are used: SL, southern limit; observers further south than this will not even see a partial eclipse; EESR, the line across the Earth where the eclipse is ending as the sun rises; P1: the first point on Earth to see the partially eclipsed Sun (at sunrise); EBSR, the line across the Earth where the eclipse is beginning as the sun rises; TMSR, totality, i.e. maximum eclipse on the umbral track, occurs at sunrise; NX, this northern crossover point (my term) exists because the August 2008 totality occurs at the top of the Earth's globe. Thus there is no natural northern limit to the partial eclipse. The northern limit of the penumbra occurs well above the Earth's surface with the practical limit being caused by the Earth's bulge, i.e. the partial eclipse becomes invisible when the Sun is below the horizon and the observer behind the terminator (i.e. local twilight/night). Looking at the previous two figures makes this clearer. Thus NX marks the point where the sunset/sunrise/start/finish lines intersect. If you like, it's the unluckiest place to be as the Sun does not quite get above the horizon while eclipsed for an observer at that point! The "loser" point might be a better term; TMSS, totality, i.e. maximum eclipse on the umbral track, occurs at sunset; EESS, the line across the Earth where the eclipse is ending as the sun sets; P4, the last point on Earth to see the partially eclipsed Sun (at sunset); EBSS, the line across the Earth where the eclipse is beginning as the sun sets. Graphic generated using *WinEclipse* by Heinz Scsibrany

part of its journey across the Earth's surface the umbra will predominantly be travelling from west to east. This can easily be verified by looking at one of the NASA/Fred Espenak global maps for the eclipse. If a total solar eclipse is being witnessed in the morning sky the shadow will be coming from behind you. The opposite applies in an afternoon eclipse, i.e. the shadow will be approaching you from the front, i.e. the direction of the Sun. Of course, these are gross oversimplifications as observers will never be looking exactly due east or west; they are more likely to be looking almost overhead if they are optimally placed on the track.

All Manner of Curved Tracks

In addition, a quick look at a typical track of the umbra's passage over the Earth will show you that eclipse paths are rarely simple tracks from west to east. They veer wildly to the north or south too. So why are these eclipse tracks so often seriously curved? What is going on? Well, the crucial point here is that the Earth is not simply rotating at perfect right angles to the Sun–Moon direction. If it was, and Sun, Moon and Earth stayed in a perfectly flat plane, the eclipse track would just be a line scored along the equator. In reality the Earth's axis is tilted with respect to its orbital plane by $23\frac{1}{2}°$ and the Moon's orbit is tilted by 5° too. Depending on whether the eclipse nodes occur in summer, winter, spring or autumn and exactly where the shadow hits the Earth (anywhere between equator and poles), a whole host of curved tracks can result. Total solar eclipse tracks in March and April will always slant from southwest to northeast across the central globe, whereas September and October tracks will slant from northwest to southeast, because of the Earth's $23\frac{1}{2}°$ axial tilt. By comparison, June and December tracks are far more likely to be parallel to the equator, at least in the central part of their tracks. If the shadow hits the Earth's surface at a very glancing angle (i.e. Sun and Moon are low in the sky) the track can be greatly widened too. The most extreme examples of this occur at sunrise/sunset regions and at high latitudes.

Anticipating the Umbra

It is a very good idea to study the umbral track path prior to setting out on an eclipse expedition because although the shadow edge is very diffuse it is certainly possible to detect the umbral shadow coming in the minutes prior to totality. If you know which direction the shadow is coming from you can easily look in that direction, and the opposite direction, as the clock counts down. That way you most definitely can detect the darkening as the shadow of the moon races towards you. If you are observing a total solar eclipse close to sunrise or sunset then the concept of which direction the shadow arrives from and departs to can be rather meaningless. At a sunrise eclipse the shadow essentially arrives from above, i.e. from space. At a sunset eclipse it departs into space after totality. Another factor to consider at such extreme cases is the extreme velocity of the shadow ground track at early morning, late evening eclipses. The speed of the umbral shadow through space will not vary all that much between lunar perigee and apogee. However, its speed relative to fixed points on the ground will depend on the angle

Fig. 3.5. As the umbra leaves the Earth's surface the umbral shadow is stretched out into a highly elongated ellipse as it glances the surface at a shallow angle. In this eclipse, i.e. 1 August 2008, the high northern track means that there is no northern limit to the penumbra, it simply merges with the terminator at the top of the Earth's globe. Graphic generated using *WinEclipse* by Heinz Scsibrany

the ground makes to the shadow axis and the lessening of the Earth's relative speed due to its direction of rotation not being perpendicular to the shadow axis. Let us take an extreme case to illustrate the point. The Sun is rising on your horizon in total eclipse, and so the ground you are standing on is parallel to the shadow axis. The shadow does not sweep across you from west to east, it descends from the sky; and the shadow hits the ground behind you at the same time as it hits the ground in front of you, so the shadow's speed along the ground is infinite. Okay, no eclipse is quite going to be like that, especially as the Earth's surface is curved, and as no one would want to view totality at precise sunrise. But I think I have illustrated the point that the speed of the Moon's shadow *across the Earth's surface* is much faster near sunrise/sunset then when the Sun is high in the sky. It might be thought that the duration of a sunrise or sunset eclipse would therefore be very short. However, because of the glancing angle the shadow cast on the ground by the umbra becomes an elongated ellipse merging with the Earth's terminator (see Fig. 3.5). The combination of high shadow speed and elongated track almost cancels out. In addition, as I mentioned earlier, the speed of the shadow across the ground at equatorial regions is substantially reduced by the fact that the Earth's rotation moves the ground beneath it at up to 1,670 km/h. This considerably extends the duration of totality at those latitudes compared with higher latitudes. At latitudes of 60° north or south the Earth's ground speed due to its rotation is halved.

A Magnificent Aircraft

The fact that the speed of the Moon's shadow across the Earth's surface can be as slow as 1,710 km/h and, therefore, within range of the (now retired) *Concorde* did not escape the attention of astronomers in the 1970s when this magnificent aircraft first entered service. Here was a plane that could keep pace with the Moon's shadow and extend the length of totality from a maximum of just over 7 min until the plane had to land and refuel (*Concorde* had a maximum range of just over 7,000 km). *Concorde*'s maximum cruising speed of 2,200 km/h (i.e. just over twice the speed of sound at 18 km altitude) could have been designed for eclipse chasing. However, staring out of tiny porthole-sized windows at totality is not in any way comparable with the normal experience of being immersed in the shadow of the Moon and being free to look around at the whole sky. Nevertheless, professional astronomers took the opportunity to mount scientific instruments inside *Concorde* as early as the 30 June 1973 total solar eclipse. The experiment was a great success with *Concorde* keeping up with the Moon's shadow over Africa for 72 min before it had to slow down prior to landing. British astronomer John Beckman and his team carried out radio observations at millimetre wavelengths which would have been impossible from ground level due to the similar sized water vapour droplets in the Earth's lower atmosphere. At *Concorde*'s cruising altitude of 18 km the water vapour problem was eliminated. In 2005, Ontario artist Don Connolly's "Racing the Moon" painting of *Concorde* 001 keeping pace with the umbra at the 1973 eclipse won first prize in the category of the commercial aviation section of *Aviation Week & Space Technology* magazine's Aviation Art Awards. The painting was originally commissioned as a cover page in Ottawa's *Citizen's Weekly* magazine. Prior to the 1960s and the launch of the first weather satellites, astronomers could only imagine what the Moon's shadow actually looked like as seen from space. However, nowadays the shadow is routinely captured by weather satellites, and even astronauts, as a black circle crossing the Earth's disc, as shown in Fig. 3.6.

Fig. 3.6. A Meteosat 8, Channel 2, image of the Moon's shadow over the Sahara on 29 March 2006 at 09.45 GMT. Image: Copyright 2006 Eumetsat

The Annular Antumbra

So far I have mainly discussed the geometry of the umbra as it hits the ground during a total solar eclipse. But what happens in an annular eclipse when the Sun appears larger than the Moon? Obviously we are not quite under the shadow of the Moon in this case, because the Sun is not totally eclipsed, but the curious reader might want to visualise what, precisely, is happening. In an annular eclipse the tip of the umbral shadow does not reach the Earth's surface. As we saw earlier, in the most extreme case the shadow falls short of the centre of the Earth by 39,400 km, or about 33,000 km short of the surface. If we draw a diagram in which we trace lines from the edges of the solar disc to the edges of the lunar disc (refer back to the right-hand diagram in Fig. 1.2), we can see that there is a region beyond the sharp tip of the umbra (where the shadow track becomes infinitesimally thin) where the lines diverge out again before they hit the Earth's surface. This is called the antumbra and where this hits the Earth's surface the silhouette of the lunar disc is seen in front of the blazing solar disc and we have an annular eclipse. The blazing ring of light means that the Sun is dangerously, blindingly and impossibly bright to look at and the corona and prominences cannot be seen visually. However, annular eclipses do have something of a novelty value, even if they are really only a special form of partial eclipse. Stray outside that antumbral shadow and the lunar silhouette breaks through the solar ring: then the annular does become a partial eclipse.

Chapter 4

The Rugged Lunar Limb

The Moon is not a perfectly smooth sphere. The mean equatorial diameter is 3,476 km and the mean polar diameter is 3,470 km. But the actual surface, although the lunar maria are relatively smooth, contains some very rugged terrain. Indeed, at the lunar limb, as viewed from the Earth, there is precious little in the way of smooth plains. Most of the maria regions are directly facing the Earth. Also, due to the so-called "librations", the limb of the Moon never presents the same jagged outline to us.

Librations

Libration is a phenomenon by which we are able to peer around the lunar limb and see, at a painfully shallow angle, the edges of the Moon's far side. In fact, we can theoretically see 59% of the lunar surface, not just 50%, by using libration tilts. As we saw earlier the Moon rotates around the Earth every 29.5 days (from new Moon to new Moon) and every 27.3 days, with respect to the stars. However, it is constantly rotating on its axis such that the same face always points towards the Earth well, almost the same face. Imagine you are talking to someone, but occasionally they lower their head, so you see more of their badly fitting wig, or raise their head so you see more of their ludicrous goatee beard. Now and again they shake their head too, so you see a bit more of one ridiculous earring and then the other. This is analogous to what librations do to our view of the Moon. Because the Moon's orbit is elliptical its angular position with respect to the Earth does not vary constantly, even though its axial rotation is always the same. The velocity of the Moon around the Earth is faster at perigee than apogee. Because of this, we can sometimes peer around either the eastern or western limbs and see almost 8° more Moon (7°54′ to be precise). This is called a libration in longitude.

There is an additional effect called diurnal libration, caused by the fact that the Earth has a radius of over 6,000 km and so, depending on whether the Moon is rising or setting (or if you are at the north or south poles for that matter) you are standing on a platform which gives you an extra ability to peer round the limb. However, most observers will view an eclipse when it is near their meridian.

The main librations in latitude (excluding travelling to the poles of the Earth) are caused by the fact that the lunar equator is tilted with respect to the lunar orbital plane, much as the Earth's equator/axis is tilted with respect to the ecliptic/ecliptic pole. Thus, as the Moon orbits the Earth, first one pole and then the other tilts by 6°41′ towards the Earth (the absolute extreme librations are actually 6°50′).

In practice, these monthly librations in latitude and longitude form a vector sum, peaking in a maximum libration effect of 10° when latitude and longitude librations peak together and swing a feature on the NE, SE, SW or NW limb towards Earth. Of course, that is not much use if the feature is in darkness, but it is highly exciting when a favourable Sun angle picks out the feature well and the sky is clear.

Watts Charts Predictions

As you can imagine, predicting just how rugged the lunar limb will be, from your location at a particular eclipse, is not a trivial task. However, a considerable amount of information is available for determining which peaks and valleys will be in silhouette at totality. For hybrid eclipses, which are on the borderline of the total/annular condition, this information is obviously crucial. But even for non-hybrid cases eclipse chasers always like to know if there is a good chance of some spectacular Baily's Beads appearing at second and third contacts. Named after Francis Baily (1774–1844), the British astronomer, these are the glistening beads that are literally the last shafts of sunlight to get through the lunar valleys at the start of totality, and the first shafts of light at the end of totality. (Incidentally, to my knowledge there is nothing to link Mr Francis Baily and the equally famous Mr Robert Francis Bailey. The latter gentleman famously "spontaneously combusted" in Lambeth, London, after drinking meths, in 1967.)

Despite the seemingly depressing and intimidating mountain (pun intended!) of work required to predict the lunar limb profile this has, essentially, already been done. In 1963, Watts of the US Naval Observatory produced a mass of data based on analysis of photographs of the lunar limb regions. Although these contained less than complete data near the poles, they still, four decades later, provide the basis for eclipse predictions. The charts are known simply as "Watts Charts". Since 1963 extra data on the polar limb regions has been added and, even today, the Tokyo National Observatory refines the Watts Charts data based on the timings of thousands of lunar occultations (stars timed disappearing behind the Moon). In 1970 and 1979, respectively, Van Flandern and Morrison proved that the lunar profile based on Watts analysis was slightly elliptical and the lunar centre implied by Watts was not exactly at the centre of the lunar mass. Morrison and Appleby, in 1981, further showed that there were tiny errors due to the average Watts radius varying with lunar librations. However, these only amounted to about 0.4 arcsec in the worst cases. This is equivalent to about 700 m on the lunar surface. Watts charts, which are provided for each eclipse typically by Fred Espenak of NASA, show a Moon in which the limb irregularities have been enormously exaggerated to allow easy interpretation of what is happening (see Fig. 4.1). If the Watts charts were reproduced without exaggeration of the limb irregularities, a 200 mm diameter Moon disc would have mountains and valleys only tenths of a millimeter above and below the perfect circle.

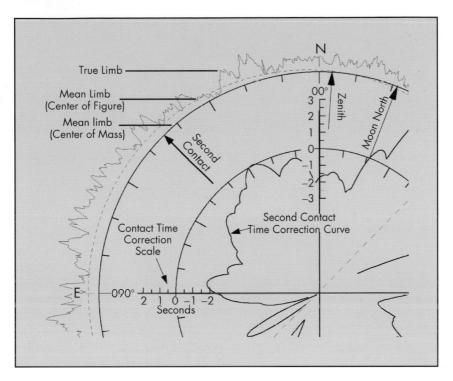

Fig. 4.1. Watts charts can be used to predict the precise timings of second and third contacts at total solar eclipses by taking into account the mountains and valleys on the lunar limb. The Watts data for the northeast quadrant of the Moon at the 29 March total solar eclipse is shown here, specifically for 10:30 U.T. and a geographic position of 28°38′47.9″N 21°53′39.9″E. The jagged outer circle is the exaggerated (by about 60 ×) rocky lunar limb as it deviates from the mean limb (*dotted line*). The next inner circle is the mean limb radius based on the lunar mass rather than the mean lunar figure. The much smaller radius solid inner circle for this quadrant marks the contact time correction zero point and the wiggling curve shows, at a large scale, the second contact time correction curve. Image: NASA 2006 Eclipse Bulletin by Fred Espenak and Jay Anderson

Hybrids and the Lunar Radius

During an eclipse, if the last or first part of the Sun disappears or reappears close to a very deep set of lunar valleys, the Baily's beads can literally sparkle, momentarily, like diamonds at the limb. The radii of the Sun and Moon during total solar eclipses are, of course, fairly similar (see Fig. 4.2). In the hybrid case the radii are essentially the same, whereas in the longest total solar eclipses the lunar radius is roughly 7% greater than the solar radius. In the former case, or at least in very short total eclipses, as the last chink of the solar photosphere disappears at the "diamond ring" position angle, Baily's beads may sparkle over a considerable arc of the lunar limb. Indeed, in the exact hybrid case, if the observer is close to the point on the track where annularity changes into totality, it may feel as if the Sun never disappeared, as the chromospheric glow on the second contact side will barely have disappeared before it reappears on the third contact side, swiftly

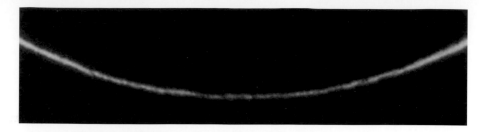

Fig. 4.2. Second contact at the 3 October 2005 Annular Eclipse, clearly showing the rugged lunar limb. Image: Damian Peach

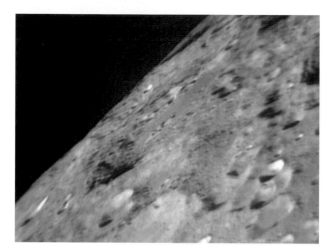

Fig. 4.3. The rugged lunar terrain and limb near the crater Demonax on 19 April 2005. Image by Damian Peach using a Celestron C 9.25 and Lumenera LU 075M camera.

followed by the emergent Baily's beads and the third contact diamond ring. As discussed earlier, the 3 October 1986 hybrid is, perhaps, the ultimate example in modern times. Predicting where an eclipse will be total along such a track requires precise attention to detail. A total eclipse is only total if the Sun is smaller than a mean lunar radius that accounts for the deepest valleys. If any photospheric light seeps through any valley at the deepest part of the eclipse, then totality has not been achieved from that viewing site. For this reason, in Fred Espenak's NASA predictions, a slightly smaller lunar radius (equal to 0.272281 of the Earth's equatorial radius) than the IAU (International Astronomical Union) lunar radius is adopted.

Short duration totalities close to solar maximum can be quite stunning, despite their brevity as, briefly, the whole Moon may be surrounded by prominences on either side. However, when the eclipses are longer, and the solar radius is much smaller than the lunar radius, Baily's beads a long way from the diamond ring point are rarer. You need a deep lunar valley, perfectly aligned, and within a few hundred kilometres on the limb circumference of the diamond ring point, to catch the smaller radius Sun. When the Sun is much smaller than the Moon, the solar surface is well below the lunar surface for points on the circumference more than 10° from the diamond ring point, as soon as second contact has taken place; likewise at third contact. In addition, all these take place very rapidly of course.

The reason that solar eclipses are so short is that the Moon slides rapidly over the face of the Sun. Put another way, the umbra moves rapidly over the Earth's surface. The Moon is moving across the solar surface, from our perspective, at up to 35 lunar km/min. So, only the deepest lunar valleys will sustain any prolonged Baily's beads. In practice, what is sometimes seen close to the critical second or third contact point is a multiple diamond ring, where one or two deep valleys might still allow the light from the blinding photosphere to travel through to the observer for several seconds around and after second contact (or before and around third contact).

In passing, even if you have little interest in using the Moon as more than an occulting disc, it is well worth studying the southern lunar limb, through a decent amateur telescope, just before or just after full Moon, when the regions near the limb show plenty of relief. The southern uplands of the Moon are spectacular, and, at the limb, on a good night, you can feel like you are in a spaceship flying over the Moon (see Fig. 4.3).

Chapter 5

Shadow Bands and Other Phenomena

Eclipses are not just about looking up into the sky. Savouring the twilight glow, the view of the planets and brightest stars, and trying to actually take in the fact that you are under the lunar shadow, are all part of the experience. Sadly, the time is all too short though and with only a matter of minutes (often panic-filled minutes if you are attempting photography too) you will always miss something. However, you should definitely try to spot the shadow bands. I have heard shadow bands described as elusive phenomena, but if you were in the Libyan Desert on 29 March 2006 you would definitely not have described them as elusive then. Shadow bands make it seem a bit like you are at the bottom of an outdoor swimming pool on a sunny day. They are dark ripples moving across the ground (or even on the side of a building, as shown in Fig. 5.1), like a collection of ghostly slithering snakes moving sideways. Shadow bands are seen just before second contact and just after third contact. In other words, they are visible when a thin sliver of Sunlight is visible, but not during totality. In Libya, where I had my best view of them, the Sun was at an altitude of 60° and the air was very clear. These two factors may well have been critical. However, another factor was that we were in a flat desert filled with light-coloured sand; so it was easy to see the not-so-subtle wiggly bands, slithering across the desert.

At this point you may ask the question "Why have shadow bands not been recorded photographically?" Well, they have certainly been recorded on video and enhanced frames show them well, but they are rather faint and the human eye and modern video recorders seem to see them better than most other forms of detector. They are also constantly moving and so a short exposure is needed to capture them, but, being faint, the more light received the better: something of a dilemma. On single video frames, their subtle nature makes them hard to distinguish. However, the human eye and brain are very good at detecting motion, even subtle and vague motion, and so seeing a video of shadow bands is far more rewarding than one freeze-frame.

Shadow Band Evidence

So what actually causes shadow bands? Well, the clue here is in the fact that they are only seen when the sunlight is coming from a narrow slit, typically less than 10 arcsec wide. Does this ring any bells? Well, stars in the night sky twinkle

Fig. 5.1. Shadow bands seen on the wall of a house in Sicily at the total solar eclipse of 22 December 1870. The eclipsed Sun would only have been at 20° altitude, which might explain why a vertical surface recorded the shadow bands so prominently.

constantly, but planets do not, and planets are, typically, tens of arcseconds across, whereas stars are point sources (at least, before their light entered the atmosphere). Of course, stars are not brilliant enough to project their twinkling onto the Earth's surface, but if they were, we would probably see a similar effect, albeit from a point, not a slit. This makes me wonder. Is it possible that the brightest historical supernovae could have cast flickering patterns onto the ground? Probably not, but I would not totally rule it out. Nevertheless, when we see twinkling stars we are, essentially, seeing shadow bands from those distant suns impacting on our retinas. However, despite my quick and confident explanation of shadow bands there are quite a few puzzling aspects to them.

Let us look, with a bit more detail, at what is actually witnessed with regard to this phenomenon. Firstly, shadow bands are not seen clearly at every total solar eclipse. This could be due to people simply not looking for them, or the surface not being conducive to observing them, or the state of the atmosphere and its ability to refract light. On quite a few solar eclipses I have been to, white bed sheets have been smuggled from eclipse chasers' hotels and laid on the ground or draped from walls to try to spot any ghostly shadow bands. German and Japanese eclipse chasers seem to be especially adept at using as many hotel fitments as possible as eclipse accessories. At the Indian eclipse of October 1995, which I observed from Fatephur Sikri, I was only a few yards from a huge white bed sheet, and yet no one spotted any shadow bands on it. (Having said that it was horribly stained, which came as no surprise to me as I knew the owners.) My gut feeling is that shadow bands, including the very ghostly ones, probably are visible at most decent total solar eclipses, but their strength and the clarity of the air play a vital role.

Ask anyone who has seen a dozen or more totalities, i.e. a real eclipse addict, and a clearer picture of what is seen emerges. Shadow bands can be seen even 2 min before or after totality. At first they are subtle and hard to see as definite bands, but as we get nearer to totality the distinctive wiggly banding becomes obvious and the spacing between the bands seems to narrow, to around 10 cm. The really long total solar eclipses, i.e. 4 min and longer seem to produce obvious shadow bands far more reliably than the short eclipses. I have already mentioned the India eclipse, where none were seen. This was only 45 s in duration. In addition, eclipses where the Sun has been at a high altitude in the sky seem to produce far more obvious shadow bands on the ground, indicating that for low altitude eclipses a bed sheet on a wall might pay dividends. A number of observers have commented that they have witnessed what appear to be more than one set of shadow bands occurring simultaneously and I have heard accounts of the separation between bands being as wide as 40 or 50 cm, or as narrow as a few centimetres. There also seems to be a general consensus of opinion that the bands are aligned lengthways with the long axis of the slit of light that produces them. In other words, if there is a narrow vertical slit of light on the lunar limb as you look up, there will be a vertical line running straight towards you as you look down. But if the slit was on the top of the Sun, you would see horizontal bands as you look down. Observers of shadow bands invariably describe them as moving. Indeed, the shadow bands I saw at the Libya 2006 eclipse looked like the ghostly souls of long departed snakes slithering from right to left across the sand, at walking pace. I have yet to hear of anyone I know seeing shadow bands at an annular eclipse, which is an important piece of evidence.

Weighing up the Evidence

So what can we make from all of this? Well, for starters, shadow bands are dark wavy lines cast by the light from a slit with a width of only a few arcseconds. This is, of course, very reminiscent of optical interference effects, i.e. the superposition of wave trains from a finite number of coherent sources, usually best demonstrated in the laboratory using lasers shining on narrow slits. The term *coherent* essentially means that the wavelengths must be identical such that there is a constant phase difference. It might be thought that these criteria would rule interference effects out; after all, the Sun emits light of all wavelengths and there is only one source, the thin sliver of sunlight. However, the crucial point in all of this is that the light is coming to us through the Earth's upper, middle and lower atmosphere which can easily refract the light such that it appears to be coming from a number of sources. In addition, although light of the same discrete wavelength is required for a perfect interference pattern, if the light does not come from coherent sources one might expect a very ghostly pattern to form on the ground: and ghostly patterns are, indeed, seen. Another way of looking at what might be happening is to just hold something like a colander up during the eclipse. The myriads of colander holes act like tiny pinhole lenses and project images of the crescent Sun on to the ground with a focal length equivalent to the colander–ground distance. Maybe part of the atmosphere is simply acting like a series of lenses and casting myriads of long, thin images of the hairline slit of the Sun onto the ground, instead of the homogeneous illumination expected from a normal healthy Sun, half a degree in diameter?

In February 1998, I was standing on the beach at Knip Bay, Curacao, in the Caribbean, where a whole host of eclipse chasers, travelling with the UK-based Explorers company, were situated. A number of prominent amateur astronomers were on the beach, including the Queen guitarist Brian May. Not far from me was the dedicated solar observer Eric Strach. I filmed a video interview with Eric and others at the eclipse site. Unlike all the other observers with video equipment his

Fig. 5.2. Eric Strach (then aged 83) at the 26 February 1998 Total Solar Eclipse at Knip Bay, Curacao, in the Caribbean. The tripod sported two cameras. The video camera, hanging below the tripod, clearly recorded shadow bands before second contact, and after third contact. Photograph by the author

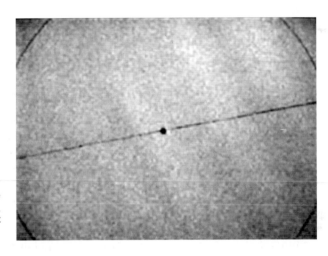

Fig. 5.3. The video recorded shadow bands of 26 February 1998, seen on white card, beneath Eric Strach's tripod. By kind permission of Eric Strach.

video camera was slung beneath the tripod and pointed down at the ground, not up at the Sun (see Fig. 5.2). It proved to be a wise move. Eric had first seen shadow bands at the Zanzibar eclipse in 1976, while looking at a white screen constructed by Liverpool Astronomical Society member Graham Broadbent. Eric eventually inherited the screen and took it along with him on future eclipse trips. At the big 1991 totality in Baja California Eric tried to video record any shadow bands present as video equipment was, at last, becoming portable and affordable. However, the experiment did not record any shadow bands on that occasion. But, seven years later, on the Curacao beach, Eric made another determined effort. A few refinements were made to the set up this time. The east–west line on the shadow band screen was carefully aligned and a one second beeper was used to generate a sound that could be used for deducing timings. Also, a Polaroid filter was used with the video camera. This time Eric hit the jackpot (Fig. 5.3). His video recording

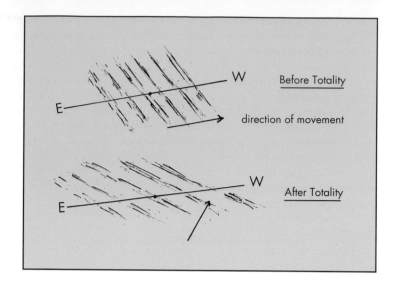

Fig. 5.4. A sketch by Eric Strach showing the westward drift of the 1998 shadow bands before totality and the southwesterly drift after totality.

clearly showed shadow bands appearing 32 s before second contact. After third contact they persisted for 27 s before their ghostly nature and the increasing sunlight washed them out. The bands seemed to travel from east to west before totality and from north–northeast to south–southwest after totality (Fig. 5.4). When the frames were frozen the bands occasionally seemed to merge. With the help of Gerard Gilligan and the loan of a frame grabber from Liverpool University, Eric established that the shadow bandwidth varied between 2.4 and 6.6 cm. In passing, it is worth mentioning that Eric Strach, like Miloslav Drückmuller, originates from Brno in the Czech Republic.

Lenses and Wind Speeds

The fact that shadow bands do seem to drift, at a relatively slow pace across the ground, is an excellent piece of evidence. Of course, being long bands, even if they are rarely dead straight, means that horizontal motion is far more noticeable than vertical motion. A vertical line moving up or down does not seem to be moving to the eye. Also, rapid motion would mean that the eye could not detect the bands: they would blur out. Zero motion would probably also mean that they were invisible, as the eye is attracted to the ghostly shadow bands simply because they are moving. Another consideration is that for the light from a very wide slit, or even the full Sun, to cause any kind of pattern on the ground due to wind motion/refraction, would require that the 10 or 20 cm diameter atmospheric cells (a scientific term for stable atmospheric regions) be very close to the ground level. Conversely, when a narrow slit or a point source is being considered, the effects of the upper atmospheric jet stream become far more relevant. The fact that the longer total solar eclipses tend to produce shadow bands more often could well be related to the fact that the dying slit of the Sun is far more curved and less slit like when the Sun and Moon are the same size, i.e. short eclipses do not produce the straight slits needed for good shadow bands. If one likens the effects of atmospheric refraction to be similar to a series of multiple lenses at all atmospheric levels, one can see that the typical speed of shadow bands across the ground (in Libya 2006 they seemed to drift at a slow walking pace) is probably related to wind speeds in the atmosphere. At first this might seem instinctively absurd because light travels at 300,000 km/s and so how can slow moving air possibly deviate such a fast light beam. However, imagine the refracting properties of the atmosphere as a series of crude floating lenses (or waves in a swimming pool) and it all makes sense. I have heard it said that bad atmospheric seeing is good for shadow band production: the more turbulent the atmosphere is, the better the shadow band is. However, I think this is too much of a generalisation. Atmospheric seeing is always poor in the daytime, with the ground radiating heat into the air, and how do you quantify "good" or "bad". As planetary observers will tell you, there are all sorts of different types of seeing, caused by the upper atmosphere jet stream being nearby, overhead, or far away and by every other level in the atmosphere.

To summarise, I think it is safe to say that shadow bands are often seen at long total solar eclipses, when the air is very clear and the ground is conducive to seeing them (e.g. a uniform sandy background) and the observers involved are specifically looking out for them. They are caused by atmospheric refraction and interference effects caused by the thin but brilliant dying slit of the Sun shining through the Earth's atmosphere. Their apparent motion on the ground is almost certainly related to local wind speeds and the direction they appear to move is related to wind direction and the position angle of the slit with respect to the observer. In general they are seen best when the solar slit is no thicker than 5 arcsec, but they have sometimes been observed when as much as 15 arcsec of slit is visible, i.e. well over 1 min before/after second/third contacts, respectively. As for recording shadow bands, well, they are faint, elusive, ghostly, and they move. So, a quality video system is needed, pointed at a plain white surface (like a hotel or cruise ship cabin bed sheet (*borrowed* for the day) or a stiff, large piece of white card) and the camera should be as sensitive as possible and with the fastest *f*-ratio lens that you can afford.

The Eclipse Micro-Climate

Ask anyone who has been on half a dozen eclipse trips and they will confirm that total solar eclipses carry their own weather along with them. Local weather forecasters always do their best to predict what is going to happen on eclipse day, but weather prediction software is not designed to cope with the effects of the Sun being blotted out in the daytime and the temperature dropping like a stone. I seem to keep mentioning my Libya 2006 experience, don't I? Well, here is another interesting snippet of data from that expedition. Keen eclipse chasers Val and Andrew White carried out a temperature test in the Sahara on that day (see Fig. 5.5). Probes measuring 1 cm under the sand, and on top of the sandy surface (in the shade), and the air shade temperature, showed temperature drops from first to second contact of 38 °C to 26.5 °C (1 cm under sand); 35 °C to 23 °C (on the sand surface); and 28 °C to 22.5 °C (air). Temperature drops of this magnitude in such

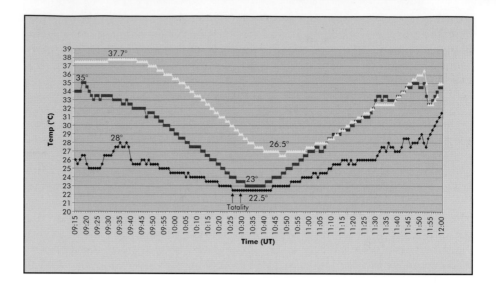

Fig. 5.5. The graph shows the results from Val and Andrew White's temperature experiment in the Libyan Sahara on 29 March 2006. The x-axis covers almost 3 h of time and the y-axis runs from 20°C to 39°C. From top to bottom the curves represent the temperature 1 cm below ground, at ground level (shade) and the air temperature (shade). See text. By kind permission of Val and Andrew White.

a short period of time can cause cloud or fog to disperse or form and wind's speeds to change noticeably, depending on what the prevailing temperature, humidity and wind speeds were before the eclipse started. If you think of the shadow of the Moon, penumbra included, being a cold spot, racing across the Earth's surface you can see that an eclipse carries its own micro-climate with it. I was at ground level in Hawaii in 1991 when what I had imagined as a sunny tropical island paradise became a cloud-filled nightmare. But as the Sun rose the cloud started to break up and we really believed that we were going to see the eclipse. Then, as the temperature took a nose dive, and with the Sun a mere sliver of a crescent, the cloud re-formed and we missed totality. Minutes later, after totality, the Sun heated everything up again and the cloud broke up. It was intensely annoying. Observers, at altitude, on Hawaii, saw the opposite effect, at totality the low cloud and fog suddenly broke up and they saw totality. Unless you are at a bone dry desert site in the middle of the day, with not a cloud in sight, then anything can happen at totality; a clear sky can cloud over and a cloudy sky can break.

Sharpening Shadows and Changing Colours

One of the most noticeable effects at a solar eclipse and one that really hits the first timer is just how weird and striking the illumination becomes as the Sun shrinks to becoming a thin crescent. Many observers will not have witnessed this phenomenon prior to their first totality because although they may have seen a few partial eclipses through the appropriate filters they may well have only seen partials where the Sun had a tiny chunk bitten out of its disc. Although others may disagree, I only really start to notice the strange illumination and the sharpening of shadows when the Sun is more than 50% obscured. I can kid myself that things look different before that, but I would not bet any money on it. Normally our daytime landscape is illuminated by a Sun that is half a degree across, i.e. it is not a point source. Shadows have a blurred edge to them because human beings have an umbra and a penumbra too. Where your silhouette is blurred the insects on the ground are seeing your body partially eclipsing the Sun. Between your body totally eclipsing the Sun and not eclipsing it at all the illumination goes from zero to 100% on the ground. The width of this penumbra can be calculated easily by

$$(\text{distance from eclipsing body to ground}) \times 0.0087,$$

where $0.0087 = $ the tangent of $0.5°$. As the Sun gets eaten up by the Moon this penumbra gets smaller, and shadows become especially sharp in one direction, i.e. the thickness of the remaining crescent. So you can hold out your hand in one direction and the fingers look sharp, and in the other direction they look blurred.

As the solar crescent becomes really thin the colours look unlike anything you have seen before. In some ways the illumination is like a twilight illumination, except the Sun is still above the horizon. The lower and sharper illumination makes the ground appear "redder" to my eyes. It is hard to define in words just how weird the lighting is at a solar eclipse but it just adds to the general feeling that you are taking part in some real-life science fiction experience. The ground illumination when the sun shrinks to a crescent always makes me think of the surface of Mars for some reason.

The Green Flash

The green flash is a phenomenon normally associated with sunset and the last tip of the Sun's disc disappearing below the horizon. It has nothing to do with total solar eclipses and yet always seems to be a popular topic of conversation on eclipse holidays. There are probably two reasons for this. Firstly, many eclipse trips involve travelling on cruise ships, where the Sun is often to be seen sinking into the water just before, during, or after the evening meal, which can infuriate the restaurant staff as all the astronomers rush out to see this phenomenon. Most people do not have a perfectly flat horizon where they live and so seeing the Sun set is a rarity. Secondly, at some eclipses, such as the 2002 December event in Australia, the Sun does set shortly after the eclipse. So there is a chance of seeing a total eclipse and the green flash too, especially from a flat desert site. The green flash is caused by atmospheric dispersion, which is greater for the shorter, bluer wavelengths than for the longer, redder wavelengths. Essentially the red Sun sets fractionally earlier than the blue Sun. So, I hear you cry "Why not call it the blue flash, or the violet flash?" Well, shorter wavelengths get scattered more too (hence our blue-coloured sky). So, in practice, the longest wavelength that gets to the observer's eye is the green wavelength.

Figure 5.6 is a dramatic combination of two events, captured by Nigel Evans, a setting solar eclipse, plus green flash. It should be stressed that it is never totally safe to look at the Sun when it is above the horizon. Even a few degrees of altitude

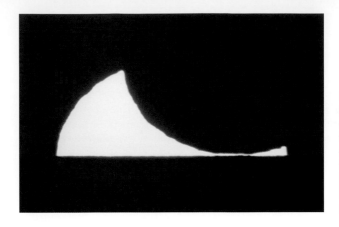

Fig. 5.6. This remarkable photograph by Nigel Evans at the 4 December 2002 total solar eclipse combines an image of the setting partial eclipse with the "green flash" (at the point where the horizontal eclipse edge is right on the horizon).

is potentially damaging to your retina. Any attempt to see the green flash should only be made during the last minute before the last part of the Sun disappears (i.e. after the Sun has half set) and even then the Sun should *never* be stared at when a significant fraction of it is above the horizon. Above all, *never ever* use binoculars or a telescope to try to see the green flash and never look through an optical camera viewfinder either. If you want to manually focus through an optical viewfinder for the green flash, then focus on a horizon object first and then move the camera onto the setting Sun without staring through the viewfinder. With modern DSLRs, just use the LCD viewfinder.

Seriously Confused Wildlife

Even the most advanced non-human animals on the planet have little ability to ponder what on Earth might be happening when the Sun's light is snuffed out in the middle of the day. With no other experiences to call on, and being blissfully unaware of how lucky they are to be situated in the path of totality, they simply assume that nighttime has arrived. Although the average eclipse chaser will have more than enough phenomena to observe during totality, the veteran will be relaxed enough to take in all the phenomena, or even have a checklist of things to savour, including the effect of totality on wildlife. In a rural location, birdsong often stops abruptly as second contact approaches; birds return to their nests or just land (or crash land!) on the ground and shut up shop for the "night". However, bats and even owls suddenly emerge, as do the mosquitoes in the hotter countries. Domestic animals may behave in a distressed manner and seek some reassurance from their owners that the world is not going to end. But perhaps the most bizarre animal behaviour of all is that of the lesser-spotted, impatient, 4×4 driver. I have seen this form of *Homo sapien*s many times at total solar eclipses. Totally oblivious of the needs of other eclipse chasers to enjoy the tranquility of the event and the subtle lighting effects, plus the darkness of totality, these strange humanoids arrive at the eclipse site at the last minute (or during totality) with headlights on full beam. Surely this is the strangest wildlife response to such an awesome event.

Chapter 6

Eclipses and Tracks 2008–2028

My intention in compiling this book was to present a highly readable explanation of total solar eclipses and advise potential eclipse chasers on how to understand, plan ahead, observe, image and enjoy future total solar eclipses. Part of the planning process involves knowing, years in advance, where total solar eclipses will occur, so some diagrams showing the tracks across the Earth's surface are necessary. It was never my intention to make this book into a set of charts for eclipse chasers, especially not when the events are many years, or decades, in the future. Firstly, the definitive guide by Fred Espenak, the NASA *Fifty Year Canon of Solar Eclipses*, contains all the global charts up to 2035. Secondly, Espenak himself issues detailed charts, well in advance of total solar eclipses, which are freely available at the NASA Eclipse Home Page at http://SunEarth.gsfc.nasa.gov/eclipse/eclipse.html.

Recently, a fine set of charts for a range of eclipses appeared in Wolfgang Held's book, *Eclipses 2005–2017*, first published in English by Floris books in 2005.

However, some detailed charts will be of interest for the reader of this book and I have endeavoured to provide them for the imminent TSEs and, in less detail, for some of the more distant "big" events, such as the total eclipses crossing the USA in 2017 and 2024.

Total and Hybrid Eclipses 2008–2027

Between 2008 and 2027 there are thirteen total and two hybrid solar eclipses. Some of the totals are long, some are short; others cross vast areas of sea, and a few cross densely populated regions. In addition, the weather prospects often vary from superb to absolutely atrocious along each track. (See Jay Anderson's web pages for more data at http://home.cc.umanitoba.ca/~jander/.) Some of the eclipses in this section and the next (annulars) classify as highly unusual, and a few follow paths that may remind experienced eclipse chasers of past events in similar parts of the world. The most hardened veterans will often note that specific eclipses are the next one in a Saros cycle to a very similar one they witnessed 18 years 11 days and 8 h earlier. In addition, they might also spot that it is quite common for rather different eclipses to occur exactly 19 years, to the calendar day, after previous eclipses. This is because 19 years is almost exactly 235 new moon intervals and only a week short of 20 eclipse years (the sun crossing the same "30–37 days wide" node twice in 346.62 days). Hence the 3 November 1994 TSE widely viewed from the Peru–Chile border region will be followed 19 calendar years later by a rather different hybrid totality off the Liberian coast on 3 November 2013.

I had originally intended to describe all the eclipses between the consecutive Saros 126 events of 1 August 2008 (#47) and 12 August 2026 (#48) in this book, but then I noticed that the 2 August 2027 totality and 26 January 2028 annular end-point were quite interesting too. A slightly modified version of Fred Espenak's NASA chart showing all total eclipse tracks from 2001 to 2025 appears in Fig. 6.1.

Another point to note when looking at these forthcoming events is that because every eclipse track crosses a huge distance on the Earth's surface there are actually some towns, villages or just positions on terra firma which can experience more than one total or annular eclipse within the space of a few years. If that place was situated at a sunny location you could even observe two major eclipses from the same site, possibly marvelling at the luck of the locals who live there. Statistically, the average place on Earth can expect to see a total solar eclipse track pass through it every 375 years or so. In his book *Mathematical Astronomy Morsels*, Jean Meeus explains that when the uplifting effect of the Earth's equator, slower shadow speed at the equator, and northern hemisphere summers occurring near aphelion are all put into the melting pot, the average time interval between total solar eclipses at a given point on Earth varies from 254 years at 80°N to 513 years at 80°S. So any land-based site enjoying two total solar eclipses, and clear skies, in the space of a decade or so is a truly wondrous place, at least in the lifetime of anyone living there. The coastal region just north of Benguala and Lobito on the eastern Angolan coast experienced two TSEs within 18 months when the 4 December 2002 followed the 21 June 2001 eclipse. The tracks of the popular European 11 August 1999 and the big African/Mediterranean eclipse of 29 March 2006 crossed in northern

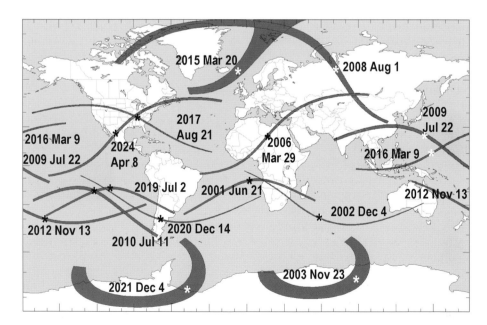

Fig. 6.1. The tracks of all total solar eclipses crossing the Earth's surface from 2001 to 2025. Note the large umbral footprint of the low altitude Arctic and Antarctic total solar eclipses. *Asterisks* mark the point of maximum eclipse. Diagram: Fred Espenak NASA/GSFC (slightly modified by the author)

Turkey, less than 7 years apart, near the town of Zara. As mentioned in the section on the 1 August 2008 eclipse, that centreline crosses the 29 March 2006 centreline near Gorno-Altaysk and Choya in southern Russia. Undoubtedly the most famous intersection in the coming years will be at Carbondale, Illinois, where the 2017 and 2024 USA centrelines cross (refer to the following sections). All four of these sites can be justifiably regarded as hallowed ground by eclipse chasers.

The Pick of the Bunch

Before we get to the specifics of each eclipse concerned, I would like, briefly, to summarise a few of the most interesting ones. Without doubt the 22 July 2009 eclipse is the real monster in the group. A mouthwatering 6 min 39 s of totality will tempt many eclipse chasers to head for the western Pacific southeast of Japan. Nevertheless, the weather prospects are not brilliant in that area in July. Just under a year later, the 2010 July eclipse is another long one at a maximum 5 min 20 s in length. But if you are happy to lose 32 s, Easter Island must be the most exciting place to see that event from. For any UK observers who were told in 1999 that the next UK total solar eclipse was in Cornwall (again) in 2090, think again. That statement may be true for mainland eclipses but the Faeroes actually host the next UK totality on 20 March 2015. Finally, the USA is really favored from 2017 to 2024, with a 2 min 40 s totality in August 2017 and then an annular and a total, 6 months apart in 2023/2024. The western USA gets an annular in 2012 too.

A couple of eclipse terms often seen in predictions may need some explanation at this point. The *magnitude* of a partial eclipse is simply the percentage of the Sun's diameter covered by the Moon at the point of maximum eclipse, i.e. the maximum linear ingress of the lunar silhouette across the solar disk, relative to its diameter. For a total or annular eclipse the *ratio* of the diameters between the Moon and the Sun is the term usually quoted, either at points along the track, or as seen from the point of maximum eclipse. Obviously, if the Moon and Sun are very similar in size (as during a hybrid eclipse) this ratio will be very close to 1.0. It will be greater than 1.0 for totals and less for annulars.

The term *Gamma* is another estimate of how "good" an eclipse is, but this time in terms of by how much the axis of the Moon's shadow misses the centre of the Earth, measured in Earth diameters. A value of 0 means the shadow axis passes straight through the Earth's centre. A value of +1 means the shadow axis just grazes the northernmost tip of the globe. A value of −1 means it just grazes the southernmost tip of the globe. In extreme and rare cases a *Gamma* fractionally larger than 1.0 can still produce a total or annular eclipse at the edges of the Earth. In this case the Moon's centre will not appear to cross directly over the centre of the Sun; so, in the case of an annular, a lopsided annulus will be seen (if, that is, anything is seen with the Sun that low down!).

Table 6.1. Maximum totality durations for total and hybrid eclipses 2008–2027

2009	2027	2010	2019	2024	2016	2012	2015	2017	2008	2026	2020	2021	2013	2023
6 min 39 s	6 min 23 s	5 min 20 s	4 min 33 s	4 min 28 s	4 min 9 s	4 min 2 s	2 min 47 s	2 min 40 s	2 min 27 s	2 min 18 s	2 min 10 s	1 min 54 s	1 min 40 s	1 min 16 s

For each of the eclipses listed below the term "Greatest", when referring to how much larger or smaller the Moon is with respect to the Sun, relates to the situation at "Greatest Eclipse", i.e. when the shadow axis is closest to passing through the Earth's centre. For total eclipses this is always very close to the point where the maximum duration of totality occurs. In addition "Shadow width at maximum" refers to the shadow width at the maximum point and *not* the maximum width.

1 August 2008 Total Eclipse

Statistics:

Saros: 47th of 72 from cycle 126.

Starts in Canada; ends in China.

Maximum duration: 2 min 27 s entering Russia.

Shadow width at maximum: 236.8 km.

Moon larger by (Greatest): 3.943%.

Best destination: China/Mongolia Gobi Desert.

If you like eclipse expeditions combined with a sunny beachside holiday, then the 1 August 2008 event might not be for you, that's assuming you buy this book within a year of publication and it is yet to happen. This eclipse (Fig. 6.2) starts at

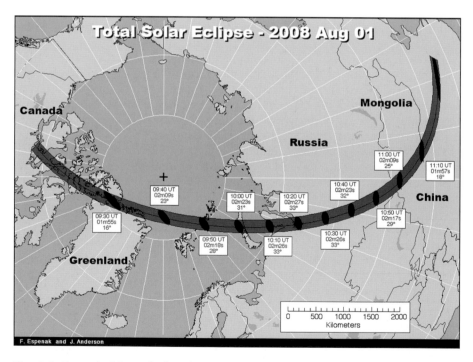

Fig. 6.2. The track of the total solar eclipse of 1 August 2008 from Canada and Greenland, across the Arctic Ocean and then through Siberia and Mongolia, ending in China. Diagram: Fred Espenak and Jay Anderson

sunrise in extreme northeastern Canada, then skirts northern Greenland and reaches the dizzy (and freezing) northern latitude of 84°N, before continuing east and curving south. The track then hits the Russian coastline close to 70°N and 70°E, not far from the stretch of water called Baydaratskaya Guba, north of the swampy west Siberian plain. It is worth remembering that the region reaches its maximum summer temperature around this time, so despite the high latitude it will be warm and mosquito infested – nice! This eclipse reaches its maximum duration of almost two and a half minutes as it crosses the Russian coastline (Fig. 6.3). But the chances of a clear sky increase rapidly as the track moves towards the Mongolia/China/Kazakhstan border, where it actually passes over the same patch of southern Russia as the track of the 29 March 2006 eclipse which many eclipse chasers observed from Libya, Egypt or Turkey. Anyone living in the villages of Gorno-Altaysk and Choya will have been under the Moon's umbral shadow twice in just over 28 months. Before that point, the southern Russian city of Novosibirsk (population 1.5 million) finds itself under the 2008 Moon's shadow at 10.45 U.T. or 16.45 local time. The city is Russia's third largest after Moscow and St Petersburg and only travelling a few miles west from Novosibirsk puts you bang on the centreline for 2 min 20 s of totality. Depending on the amount of cloud cover and wind direction the afternoon temperature at Novosibirsk on 1 August can be anywhere between 20 and 35°C, but it will not be cold. Residents of the city should not experience any telescope shortage as the factory that produces the well-known TAL telescopes is situated in Novosibirsk. The instrument making company began manufacturing telescopes for amateur astronomers in 1980

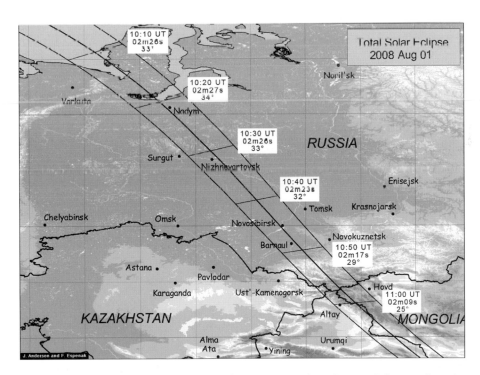

Fig. 6.3. The track of the total solar eclipse of 1 August 2008 through Russia/Siberia and into the China/Mongolia border region. Diagram: Jay Anderson and Fred Espenak

after an initiative by local amateur astronomer Leonid Sikoruk. So, if you chose Novosibirsk to see the 2008 eclipse, a trip to the TAL factory might be a good idea.

At 10.55 U.T. the eclipse leaves Russia, briefly skirts the eastern Kazakhstan border and then skirts the Mongolia/China border and western Gobi desert for the next 500 miles before proceeding deeper into northwestern China (Fig. 6.4). Much of the Gobi Desert is hilly or mountainous but between Altay and Hami, at 45°N, the Dzungarian Gobi, inside the Chinese border, is relatively low altitude. Daytime Gobi temperatures at sea level in August typically peak anywhere between 20 and 38°C. But at altitude, and in a northerly breeze, temperatures can be much cooler. However, being a desert region this area offers the most likely prospects of a clear afternoon sky in August; around 60 or 70%, although the Sun will be just below 20° in altitude. The rest of the track is much cloudier in a typical year. Those eclipse chasers considering a long holiday in China may like to consider the fact that the Beijing Olympics are scheduled to run from 8 to 24 August. So if you are into eclipses and sport and have a month to spare, this could be the holiday of a lifetime. The 2008 total solar eclipse ends at sunset near Zhengzou, 800 km east of the Himalayan foothills, half that distance west of the Yellow and East China Seas, and 600 km south of Beijing itself. If sunset were a few minutes later the track would pass just north of Shanghai, but as it is the eclipse does not reach the east China coast.

During totality four planets and a first magnitude star stretch eastwards along the ecliptic for 40° from the Sun. Closest in are Mercury (4°), Venus (16°) and Regulus (20°), with Saturn (30°) and Mars (40°) also detectable. In addition the well-known "Beehive" star cluster M44 will be only 2° to the Sun's northwest

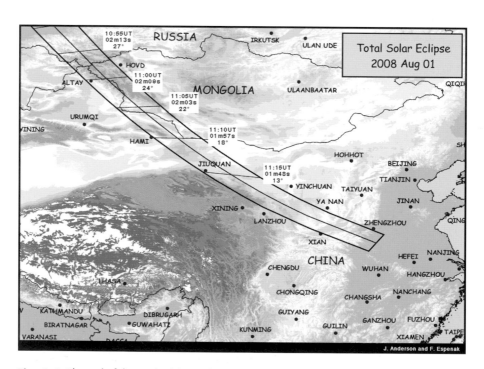

Fig. 6.4. The end of the track of the total solar eclipse of 1 August 2008 through Russia/Siberia and into the China/Mongolia border region. Diagram: Jay Anderson and Fred Espenak

during totality. It is quite possible that some of the cluster's brighter stars (sixth magnitude) may be recorded on long outer corona exposures.

The altitude of the Sun above the horizon is always less than 34° for this eclipse, wherever you go, which will be a blessing to those with an artistic eye (who like a scenic backdrop to an eclipse) and those who hate those near-zenithal neck-cricking eclipses, where tripods and home-made telescope mountings are close to falling over. An extra bonus is the fact that the later stages of the track offer a good chance of clear skies on a land-based site in the civilised part of the day, i.e. the afternoon. Those eclipse chasers who have experienced many pre-dawn bus rides to get to early morning eclipse sites will know how good that sounds. With the Sun now past the predicted minimum activity levels of 2006 we may expect to see a corona which is less east–west dominated than at exact solar minimum and maybe a good prominence or two will put in an appearance, if we are lucky. The previous eclipse in this Saros was the 22 July 1990 event which many European eclipse chasers will remember for their rather depressing cloudy sunrise experience in Finland. Fans of the British TV astronomer and amazingly prolific author Patrick Moore (who is not a fan of Patrick's?) might care to read Chap. 14 of his autobiography, "80 Not Out", before heading for Siberia. Patrick's experience of the 40 s duration 22 September 1968 totality from Yurgamysh makes entertaining reading. Of course, that era was deep in the cold war years of the 1960s. Plenty of options exist for travel to the Gobi Desert for the 2008 eclipse and a journey on the Trans-Siberian Railway from Moscow to Novosibirsk must be a big attraction. In terms of cloud cover, Jay Anderson's research for July and August along the track shows that the probability of cloud cover is poor over the early part of the track and peaks at over 90% near Spitsbergen but improves rapidly as the track curves towards the Russian border with Mongolia and China. At Novosibirsk, the probability of cloud drops to around 50% and the lowest probability of cloud occurs at a longitude of around 97°E, 42°N close to Mingshui and Gongpoquan in China, just south of the Mongolia border. A cloud cover probability of only 30% is likely at that point, although the weather on eclipse day has never obeyed statistics.

22 July 2009 Total Eclipse

Statistics:

Saros: 37th of 71 from cycle 136.
Starts in India; ends in S. Pacific.
Maximum duration: 6 min 39 s SE of Japan.
Shadow width at maximum: 258.5 km.
Moon larger by (Greatest): 7.991%.
Best destination: W. Pacific, SE of Japan (for longest duration).

By any definition this total eclipse is a full-blown monster. It is yet another super-long eclipse from the famous Saros cycle 136; the same cycle that spawned the twentieth century's big totalities on 20 June 1955 (7 min 8 s), 30 June 1973 (7 min 4 s) and 11 July 1991 (6 min 53 s). Why should this Saros spawn all the really long eclipses? Well, simply because in recent decades its eclipses have been in June and July, when the Sun is smallest in size, and the Moon has been close to perigee too.

Added to this is the fact that the eclipses of this Saros are near the middle of their working life (this one is the 37th of 71) and so occur well away from the polar regions. Saros 136 eclipses are past their best though and this 6 min 38 s monster is the longest total solar eclipse of the twenty-first century. The 38th eclipse of this Saros, on 2 August 2027 will only be 6 min 23 s at its peak (Egypt). In fact the 2009 eclipse offers the longest period of totality until 13 June 2132. Unfortunately, nowhere along the track guarantees a high probability of clear skies, although being on a mobile cruise ship southeast of Japan offers the best prospects, combined with the flexibility to move to the most likely clear patch in the days before 22 July. Through China and out into the Pacific Ocean the probability of seeing totality is no better than 50%; a dismal situation for such a major eclipse. Choosing the fastest cruise ship with the best weather radar may be the optimum strategy for this one.

The 22 July 2009 eclipse starts, at sunrise, on the west coast of India (Fig. 6.5). Even at this starting point more than 3 min of totality is on offer. The umbra rapidly moves east–northeast, heading for the rather confusing northeastern corner of India where the borders of Nepal, Bhutan and Bangladesh all but cut off the Indian Territory, but not quite. Mount Everest, on the Nepal/Tibet border, is just 100 km north of the track edge. The Moon's shadow passes over the historic town of Darjeeling, in the northernmost tip of Indian Territory, famous for its tea plantations, Himalayan railway, western-style public schools and missing manhole covers (avoid these in the dark). However, the probability of clear skies there in July is very slim. (I visited Darjeeling in November 1998 in an attempt to view the 1998 Leonid meteors. It was supposed to have clear skies, but a typhoon followed

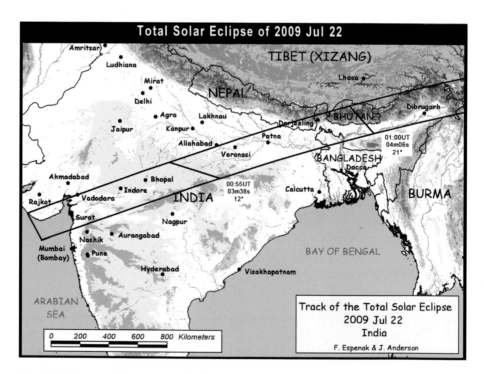

Fig. 6.5. The start of the umbral track of the 22 July 2009 total solar eclipse as it passes over India and the edges of Nepal, Bhutan and Bangladesh. Diagram: Fred Espenak and Jay Anderson

us there and all we saw was torrential rain!) After leaving northeast India and the extreme northern Burmese boundary the eclipse track heads east across China (Fig. 6.6) where the disappointed residents of Shanghai (see the 2008 eclipse assessment) make up for missing the last eclipse. This time the track passes right over the famous port and, if skies are clear, the population will enjoy over 5 min of totality. By travelling 100 km southwest, to Jiaxing, this increases to a stonking 5 min 56 s. If you really do not like cruise ships then Shanghai/Jiaxing is pretty much the last piece of mainland terra firma that you can view the eclipse from, although there are a few islands that the shadow crosses over as the maximum phase approaches. In addition, some of the modern cruise ships are big enough to class as islands in their own right. After Shanghai, the shadow speeds across the East China Sea and then encounters some of the southernmost Japanese islands (Fig. 6.7). Around 130°E of Greenwich a whole host of small Japanese islands with names ending in Jima or Shima straddle the umbral shadow from north to south, offering up to 6 min 20 s of totality. Best placed, fractionally north of the centre line, is the small island of Suwanose-jima. But, to the north, Nakano-shima, and to the south, Akuseki-jima, also offer 6 min of totality. Further east (1,200 km) the umbra passes just south of Ogasawara-shoto (the Bonin Islands) and right across the tiny island of Kitaio-jima. Here, the period of totality peaks at 6 min 39 s with the Sun virtually at the zenith (85° altitude!). Plan to lie on a sunbed with a comfortable pillow if you are on this part of the track. As the umbra speeds across the Pacific Ocean, it only encounters one other chain of small islands on its journey. These are the Gilbert Islands, renamed the Kiribati Islands after they became

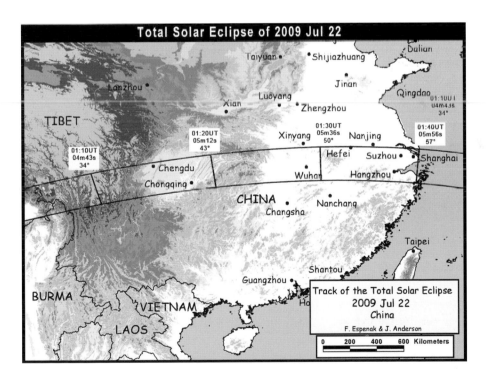

Fig. 6.6. The umbral track of the 22 July 2009 total solar eclipse across the width of China. Diagram: Fred Espenak and Jay Anderson

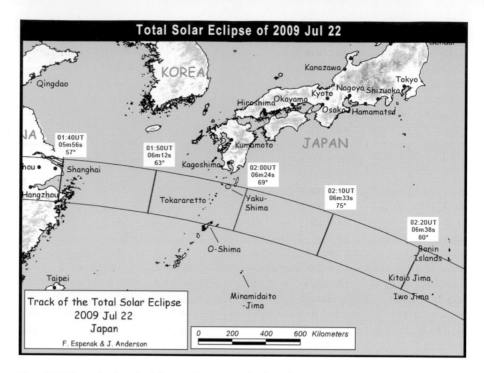

Fig. 6.7. The umbral track of the 22 July 2009 total solar eclipse at the point where the eclipse peaks in duration, i.e. as it leaves China and crosses to the south of Japan. Diagram: Fred Espenak and Jay Anderson

independent from the UK in 1979, and they lie roughly 1,700 km north of Fiji. The island of Marakei is nearest to the centreline and will enjoy 4 min 44 s under the umbra. At this longitude (173°E) the eclipse occurs in the late afternoon and the Sun will be at a more comfortable altitude of 38° above the west horizon. Coincidentally, the Gilbert Islands will experience an annular eclipse 4 years later, on 10 May 2013. Some 600 km east of Samoa the speeding umbra will finally leave the Earth's surface at sunset. The longest total solar eclipse left in our lifetimes will then be over, unless anyone seeing it lives to at least 123 years of age.

11 July 2010 Total Eclipse

Statistics:

Saros: 27th of 76 from cycle 146.
Starts in SW Pacific; ends in Argentina.
Maximum duration: 5 min 20 s in S. Pacific.
Shadow width at maximum: 258.7 km.
Moon larger by (Greatest): 5.804%.
Best destination: S. Pacific/Easter Island.

In 2010, for the third successive "northern hemisphere summer" in a row, we have a good total solar eclipse to head for; a fairly unusual situation. However, this eclipse is confined purely to the far southern hemisphere and, at first glance, only touches mainland territory at the very sunset end of the track, where it crosses Chile and ends in Argentina (see the global map at the start of this chapter). However, take another look and you see that the track passes over the famous "Easter Island" (Fig. 6.8) as well as several small islands in French Polynesia in the Hikueru and Amanu atoll regions. At 5 min 20 s this is another long eclipse and we have to wait until 2 August 2027 before there is one to beat it in duration terms (the next Saros 136 monster). Unfortunately, the southern Pacific Ocean can be an especially cloudy place in July and so although Easter Island is a very tempting place to be during totality (amongst all those amazing giant statues) a more sensible approach would be to visit Easter Island as part of the holiday and be on a cruise ship with plenty of manoeuvrability. The ability to steam towards the most likely clear patch in the days before the eclipse, when forecasts become more reliable, is essential to maximise your chances. According to Jay Anderson's weather research the prospects along the most northerly part of the track, just before maximum eclipse, are the best. Specifically, in the French Polynesian Island region there is a 50% chance of cloud cover. (Less if you are on a ship that can steam towards a big gap!) Things get dire as the track approaches Easter Island and South America though. Here we are looking at a 70–80% chance of cloud cover. It is easy to say "maybe we will be lucky" in these situations, but, as a victim of Hawaii in 1991 my mind immediately looks for the word "Desert" on any eclipse track and statistics of 90% in my favour.

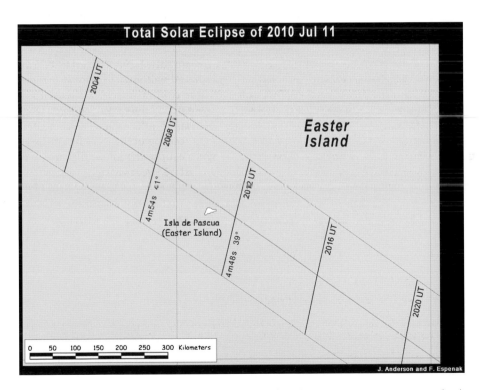

Fig. 6.8. The umbral track of the 11 July 2010 total solar eclipse as it crosses over Easter Island. Diagram: Jay Anderson and Fred Espenak

If you simply like foreign travel though, watching the eclipse from the end of the track, at Chile (Fig. 6.9), should not be overlooked. From a European perspective Chile/South America is not too difficult to travel to (typically via Schiphol/Amsterdam airport in the Netherlands) and a far less daunting journey than heading for New Zealand, which is not too far from the start of the track. In terms of longitude Chile is only 5 h behind London, whereas New Zealand is the worst possible jet-lag scenario, i.e. 12 h different from the Greenwich meridian. Nevertheless, we are talking about the extreme southern tip of Chile here, at the same latitude as the Falkands and not far from Cape Horn! The big problem from Chile is the Sun's altitude. At best (near the west coast) the 2 min 56 s of totality will only be 4° above the Pacific horizon. It would certainly be a surreal and picturesque sight, reflected in the sea horizon, but the chances of seeing it when that low down are very small indeed: even a few clouds in the distance would hide it. From Argentina the altitude sinks to 0°. Conversely, from a cruise ship in the south Pacific, positioned at 19°46.5′S and 121°51.0′W, the full 5 min 20 s can potentially be seen, weather permitting, with the Sun at a fairly comfortable altitude of 47° at midday. Earlier along the track, the French Polynesia regions east of Tahiti will see a minute less of totality with the Sun in the morning sky and at an altitude of 30°. Later along the track the aforementioned Easter Islands will see a 4 min 48 s eclipse in the afternoon, and at an altitude of 40°. All in all a cruise ship in the French Polynesia or Easter Island vicinity sounds like a good expedition, depending on your priorities. As always,

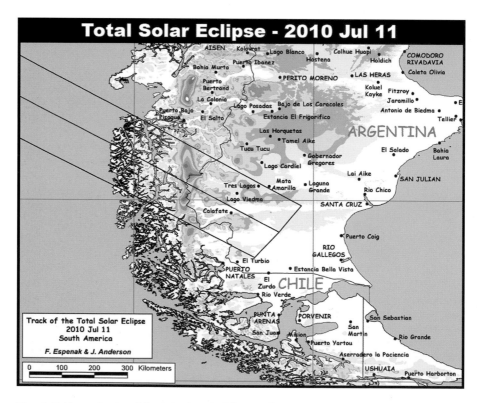

Fig. 6.9. The end point of the umbral track of the 11 July 2010 total solar eclipse as it passes over Chile and ends, at sunset, in Argentina. Diagram: Fred Espenak and Jay Anderson

Mercury and Venus will be the easiest planets to spot at totality, and Mars, Saturn and Regulus continue the eastward curve of bright objects.

13 November 2012 Total Eclipse

Statistics:

Saros: 45th of 72 from cycle 133.
Starts in Australia; ends in SE Pacific.
Maximum duration: 4 min 2 s in S. Pacific.
Shadow width at maximum: 179.0 km.
Moon larger by (Greatest): 5.005%.
Best destination: Queensland, Australia for least cloud.

With 28 months separating the 2010 and 2012 totalities, diehard eclipse chasers will be keen to resume their passion and get their next coronal fix. However, I suspect the 6 June 2012 transit of Venus across the face of the Sun will lure many umbraphiles even if the disc of Venus falls way short of totally eclipsing the solar disc! The Venus transit will be best viewed in the western Pacific, but some Europeans may glimpse the end of the event at sunrise. Back to the main plot: the 2012 total solar eclipse is a bit like the 2010 totality in reverse (see the global map at the start of the chapter). The mainland section is at the start, not the end, in northern Australia this time, but the main track, once again, crosses the southern Pacific Ocean. This time it passes roughly 2,000 km south of Easter Island before curving north to cross the 2010 track. But in 2012 the track does not reach the South American mainland. Just over 4 min of totality are on offer at 39°57.6′S and 161°17.9′W, but, without doubt most eclipse chasers will be seriously looking at the sunrise east coast of Australia (Fig. 6.10), despite only 2 min and 6 s being available (just north of Cairns) and the Sun being only 14° above the Coral Sea. With clear skies Eastern Australia will also be a good place to view the transit of

Fig. 6.10. The crossing Australian ground tracks of the 13/14 November 2012 total solar eclipse (starting near Maningrida and heading southeast) and, 6 months later, the 10 May 2013 annular solar eclipse (starting in the Great Sandy Desert and heading northeast). Diagram by the author using *WinEclipse* by Heinz Scsibrany

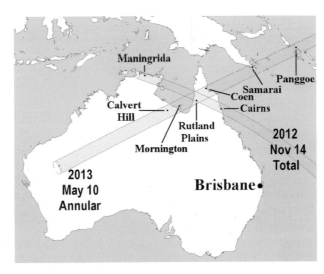

Venus some 5 months earlier. Scuba diving eclipse chasers (and there have been quite a few, including this author) will also note the proximity of the Great Barrier Reef. If you opted not to go to the 2002 December totality in the southern Outback, because of its short duration, you get a second excuse to see Australia 10 years later. Remarkably, Northern Queensland, and the zone lying roughly between Rutland Plains and Old Strathgordon will experience an annular eclipse only 6 months later! If you plan observing this totality from Australia, then the coastal region near Cairns is definitely the place to be, unless they are cloudy or foggy on the day. The Sun will only just be above the horizon at Arnhem land in the Northern Territory region, which is an aboriginal reserve anyway. On the east coast of Queensland, just north of Cairns, the Sun is just high enough to get an undistorted view, and the hour between sunrise and totality is just enough to watch the initial partial phases. First contact is at sunrise, although the solar disc will be distorted by refraction, so the early partial phases will be very tricky to observe. The big question with this eclipse is: do you want to see a 2 min and 6 s totality from Australia, or 4 min and 2 s from a cruise ship in the Pacific? If you usually fiddle about with camera equipment, a 4 min eclipse can seem like 30 s anyway. Maybe this is an eclipse to just sit back and savour visually, from a deckchair on the beach north of Cairns? This region offers the best clear sky prospects, namely around 50%. However, as with any low-altitude eclipse even a few clouds on the horizon can wreck the view. With the Sun so low down from Australia only Venus will be obvious. It will be well above and to the far left (35° western elongation) of the corona. Saturn is closer, at half the distance, but at magnitude 0.6 will definitely not be that obvious at an altitude of 14°. With the Sun now at the maximum of its solar activity this totality has every chance of showing considerable activity, with plenty of sunspots on the disc, some decent prominences with a bit of luck, and a more symmetrical corona than we see at the solar minimum eclipses. The corona should be obvious all around the disc, not just at the east and west position angles.

3 November 2013 Hybrid Eclipse

Statistics:

Saros: 24th of 73 from cycle 143.
Starts in W. Atlantic; ends in Kenya.
Maximum duration: 1 min 40 s of totality near Liberia.
Shadow width at maximum: 57.5 km.
Moon larger by (Greatest): 1.588%.
Best destination: Off the Liberian coast?

There are only two hybrid eclipses between 2008 and 2024, and this is the first one. It is the first hybrid eclipse since 8 April 2005. If you jumped over the first few chapters I will just re-iterate that a hybrid eclipse, sometimes called an annular total eclipse, is one where the Moon is so close in size to the Sun that the eclipse is annular at the world track edges (one or both), and total in the middle. The observer is raised up as the Earth rotates and, at local noon, the Moon can look almost 2% bigger in the most extreme cases. Not surprisingly, the totality phases of hybrid eclipses are never long. Hybrid eclipses with a totality phase longer than one and a half minutes are very rare indeed, but this is one of them, and if you regard it as simply a short duration total eclipse, you will not be far wrong. There is only the briefest annular phase, i.e. at sunrise in the western Atlantic Ocean. The whole of the rest of the track experiences the umbral shadow. To enjoy the full 100 s of totality though you need to be on a ship just off the coast of Liberia, where the Sun will be 70° above the horizon. Unfortunately there are no deserts along the track and even off Liberia, a relatively good site from this viewpoint, the probability of cloud cover is high, about 60%. To me this eclipse has similarities to the famous 1973 total solar eclipse when the good ship *Monte Umbe* sailed from Liverpool in the UK to observe that huge seven minute totality off the coast of Mauritania. One of the highlights of that cruise trip was the TV astronomer Patrick Moore's trousers split-ting on the deck. More pictures were taken of that event than of Totality! Patrick's last night composition and performance of the song "Nouadhibou", describing the evil smelling, fly-ridden godforsaken sandswept dump in which they briefly stopped on the way back, has gone down in the annals of eclipse chasing! As well as being manoeuvrable, a cruise ship is safer and more civilised in this part of the world and the waters off Liberia offer the best prospects and longest duration for this 2013 event. The track of this eclipse starts in the Atlantic, as an annular event, just off the eastern US seaboard. A partial eclipse will briefly be seen at sunrise from the whole eastern US coastal region. The eclipse becomes total almost imme-diately and sweeps across thousands of kilometres of uninhabited ocean before neatly sliding under the bulge of equatorial Africa. Refer to Fig. 6.11 which also

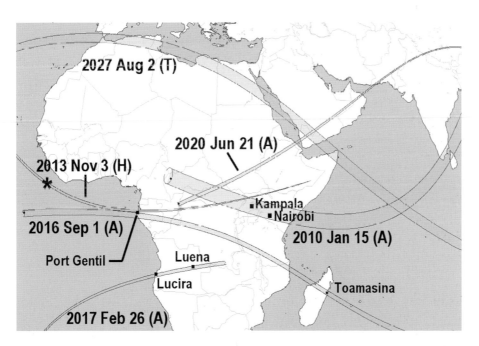

Fig. 6.11. In the space of just over 7 years (2010–2017) one hybrid eclipse and three annular eclipses cross parts of the African continent. A further annular starts in Africa in 2020 and a total skirts the northern edge of the continent in 2027. As described in the text the 3 November 2013 hybrid event is the best one and peaks as a total eclipse just off the Liberian coast. Diagram by the author using *WinEclipse* by Heinz Scsibrany

shows the annular tracks across Africa in 2010, 2016, 2017 and 2020 and a total eclipse in 2027. The track does not actually hit the African coast until well after its maximum point. At 13 h 50 min G.M.T. the eclipse enters Gabon, just north of Port Gentil (the 2016 Annular enters Africa just south of Port Gentil!). Already the period of totality has dropped to 1 min 10 s, and as the umbra enters the Congo it drops below 1 min. The final stages of this eclipse take it over Uganda, Kenya and Ethiopia, where it ends, at sunset, on the Ethiopia/Somalia border region. With the Sun still around its maximum activity period, a total solar eclipse, where the Moon is not much smaller, offers a slim prospect of a lunar disc surrounded by pink prominences . . . we can but hope!

20 March 2015 Total Eclipse

Statistics:

Saros: 61st of 71 from cycle 120.
Starts in N. Atlantic; ends at North Pole.
Maximum duration: 2 min 47 s nr Faeroe Isles.
Shadow width at maximum: 462.0 km.
Moon larger by (Greatest): 4.455%.
Best destination: Spitsbergen?

As I mentioned earlier in this book, if you were around in Cornwall in 1999, sitting under a huge bank of cloud, you may well recall the cheerful and optimistic comments that went something like: "Well, that's the last British total eclipse until 23 September 2090, and we'll all be dead by then . . . bummer". That 2090 total eclipse will be over Cornwall too, but there are two others between 1999 and 2090 that are still within British waters. One just skirts the Channel Islands in 2081 (3 September), but the 2015 totality passes over the Faeroe Islands, off Scotland's north coast, midway between Scotland and Iceland (see Fig. 6.12). The weather at that time of year, despite being 3 months past the Winter solstice, is often bitterly cold and if you fall in the sea at 62°N, they will be pulling your rigid corpse out after a few minutes! Add to that the fact that it will almost certainly be cloudy, if not pouring with rain, and the Faeroes' prospects do not look brilliant. On the bright side though, it is not far, or expensive to travel to if you are Scottish, and the Scots are renowned for paying the minimum for everything they purchase. The Faeroes are not bang on the centreline of this eclipse though, although they are the first major island that the track hits after leaving the waters near Newfoundland at Sunrise. (The umbra just passes over Rockall too, but being a 30 m wide rock inhabited by thousands of birds, it is not much of a holiday destination!) Two minutes and seventeen seconds of totality is potentially available from the main Faeroe town of Tórshavn, but 2 min 48 s are available 150 km to the west, if you like choppy seas and seasickness! The Sun will be at an altitude of 20°. (The duration of the Cornwall totality, 16 years earlier, was only 2 min and 2 s.) After the Faeroes the umbra speeds even further north and passes over the frozen Norwegian islands of Svalbard, in particular the main island Spitsbergen (Fig. 6.13). The main town Longyearbyen is on the centreline and its airport, at 78° north is the world's most northerly fully operational airport with daily flights to Tromso, Oslo and Braathen. The flight to

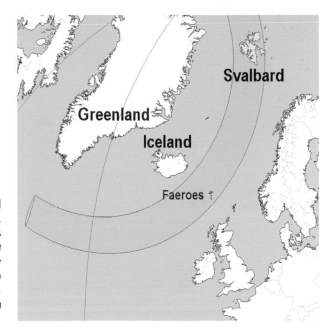

Fig. 6.12. The next total solar eclipse in UK waters, that of 20 March 2015, does not hit the mainland but crosses over the cloudy Faeroe islands and on towards the Svalbard islands. Diagram by the author using *WinEclipse* by Heinz Scsibrany

Fig. 6.13. The umbral track of 20 March 2015 as it passes over the Svalbard archipelago and the frozen island of Spitsbergen. Longyearbyen is in the centre of the umbral track. Diagram by the author using *WinEclipse* by Heinz Scsibrany

Oslo takes 3 h. Spitsbergen may be a name familiar to some amateur astronomers who are keen Aurora watchers as it is ideally placed for such activities, except in summer when the Sun never sets. Even at this time, a few years after solar maximum, the site provides a good chance of seeing spectacular winter auroral activity. A full-time Auroral station at Adventalen, just outside Longyearbyen, has been operating since 1978. The island has quite a few hungry polar bears too and there have been a few fatalities over the years. Nevertheless, the island has quite a healthy northern lights tourist industry for Aurora fans. Unfortunately, even as early as 20 March, the nights are short and not totally dark. When you are at 78° north it is difficult to get the Sun well below the horizon unless you are in the winter months. The Sun is below the horizon by 8° or more (the twilight criteria used by the local Aurora station) from 19.56 to 02.19 U.T. on 20 March and goes no deeper than 12° down. One week earlier, on 13 March, the corresponding 8° figures are 18.53 to 03.24 U.T. with the Sun 15° below the north horizon just after local midnight (23.00 U.T.). Of course, the Moon will be new and completely out of the night-time sky in the days before and after eclipse day, 20 March.

The Radisson SAS Polar Hotel at Longyearbyen is one of the few hotels on the island. Spitsbergen in March is not for anyone who craves warm weather! In that month one can expect a daytime temperature peaking at −15°C and a nighttime low of −23°C. High wind speeds and wind chill temperatures of −40 or −50°C are quite common in March too! So if you have camera equipment with you there is a strong probability that it will fail totally. Batteries do *not* like temperatures below freezing point, and certainly not *that* low. A system powered from mains electricity will be much more reliable. Spitsbergen is only 1,300 km from the north pole and it is a cloudy place in March too (though not as cloudy as the Faeroes). But it is a spectacular frozen and rocky landscape and the sight of a totally eclipsed Sun above that scenery, for 2 min and 30 s, would be absolutely awesome. More so, because of the low solar altitude of 11°; an artist's surreal landscape dream. Make sure there are no mountains to your southwest or you will not see the eclipse at all! After Spitsbergen the umbra continues literally to the very door step of the North Pole, leaving the Earth's surface at 88°N, just before it arrives there at sunset. Due to the high northern latitudes that this eclipse travels over, the shadow has a very pronounced ellipticity, stretching out to a northwest–southeast footprint of 462 km width at the maximum eclipse point. Mercury, Venus and Mars will all come out to play at totality, but if you are lucky enough to see the eclipse from Spitsbergen you will not remember them! Pack some serious, specialist, winter clothing for this one and approach it as you would a trip up Everest! It is not impossible that a cruise line might offer a trip to the umbra in the North Atlantic for this one, which would be a warmer, but less spectacular option.

9 March 2016 Total Eclipse

Statistics:

Saros: 52nd of 73 from cycle 130.
Starts in Indian Ocean; ends E. Pacific.
Maximum duration: 4 min 9 s in W. Pacific.
Shadow width at maximum: 155.0 km.
Moon larger by (Greatest): 4.499%.
Best destination: Mobile cruise liner!

One year after the freezing cold 2015 eclipse there is a *much* warmer prospect in store, with a full 4 min and 9 s on offer, and the morning part of the track crossing the Indonesian islands. (See the second near-horizontal track from the top in Fig. 6.14.) Once again, the major part of the track passes over the Pacific Ocean, although as that mass of water is so enormous this can hardly come as a big surprise. The eclipse begins, at sunrise (as always), in the Indian Ocean, and rapidly heads east until it crosses the Kepulauan Mentawai region islands Pagai Utara and Pagai Selatan just prior to arriving at the east coast of Sumatra. As the umbra crosses central Sumatra totality exceeds 2 min on the centreline. The track then crosses the southern part of Bangka island, the whole of Belitung island (although the north is favoured) and then across the Java sea to the west coast of Borneo, where, presumably, the wild man will be observing the eclipse (unless he has resumed his job as an airline baggage handler). After Borneo the eclipse flies over the Java Sea again and crosses the northern half of Sulawesi, the Molucca Sea and the central part of the smaller island Halmahera. In central Halmahera the eclipse duration reaches 3 min and 19 s, only 50 s short of the maximum for this eclipse. After this point the umbral track curves north, into cooler waters as it speeds across the Pacific Ocean. The maximum 4 min 9 s point occurs at 10°6.5′N and 148°50.2′E. This far into the Pacific the cloud cover is considerably better than in the tropical Indonesian region, so, once again, a cruise ship looks like a good option. The sunset end of this eclipse occurs roughly midway on a line between Hawaii and San Francisco, approximately 1,500 km north–northeast of Hawaii, so a cruise ship sailing from Hawaii to the nearest part of the track to the north is a possible option (assuming the travel companies offer this in 2016).

In 2016 the Sun will be back at its minimum state of solar activity, so one can expect quite a few long equatorial streamers but with a minimum of prominence activity, unless we are very lucky. Mercury and Venus will be easy targets to the west of the Sun, shining at magnitudes –0.6 and –3.9, respectively.

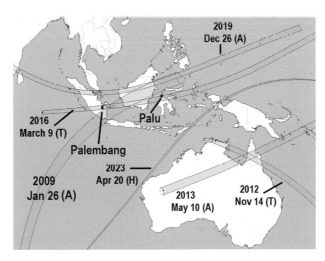

Fig. 6.14. Two total, three annular and one hybrid eclipse cross the Malaysia, Indonesia and Australia regions between 2009 and 2023. The Australian events were shown in more detail in Fig. 6.10. Diagram by the author using *WinEclipse* by Heinz Scsibrany

21 August 2017 Total Eclipse

Statistics:

Saros: 22nd of 77 from cycle 145.
Traverses the USA from the west coast to the east coast.
Maximum duration: 2 min 40 s in Eastern USA.
Shadow width at maximum: 114.8 km.
Moon larger by (Greatest): 3.059%.
Best destination: West & Central USA favoured.

The first total solar eclipse to cross the USA since 26 February 1979 (that one crossed Washington/Oregon, Idaho and Montana) occurs on 21 August 2017. It seems odd that a country as big as the USA should have missed out for the intervening 38 years, but in 7 years time, on 8 April 2024 the USA gets another one, as well as two annulars in just over 11 years in 2012 and 2023. The tracks of these events, plus one other crossing North America, are shown in Fig. 6.15. Between 1900 and 1999 the US mainland experienced 11 total solar eclipses in 1900, 1918, 1923, 1925, 1932, 1945, 1954, 1959 (just!), 1963, 1970 and 1979. Between 2000 and 2100 it will experience eight more, in 2017, 2024, 2044, 2045, 2052, 2078, 2079 and 2099. So, one TSE per decade is about the USA norm. The tracks for 2052 and 2078 are almost parallel and cross at the point where the extreme southeastern Texas border joins Mexico on the Gulf Coast. Remarkably, a third umbral track crosses this point too, in 2017, although as it is essentially across Mexico it is not included in the above. (The track

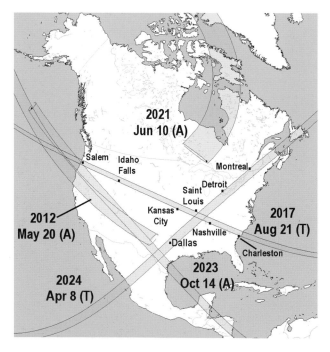

Fig. 6.15. Three annular eclipses and two total eclipses cross North America between 2012 and 2024. The 2017 and 2024 total eclipses, crossing the USA will attract millions of first-time eclipse chasers. Diagram by the author using *WinEclipse* by Heinz Scsibrany

of the 2023 annular leaves the USA just north of this point too!) Of all the mainland USA crossing tracks in the 200 year period from 1900 to 2100, only those of 1918, 2017, 2024, 2045, 2078 (arguably) and 2099 cross a large swath of the USA and the first four are definitely the best. So the 2017–2024 time span is a really exceptional one if you are a USA eclipse chaser, even without the 2023 annular eclipse. Do the 2017 and 2024 paths of totality cross I hear you ask. Yes Indeedy! The tracks are literally at right angles to each other and cross near to Carbondale Illinois, southeast of Saint Louis near the Missouri/Illinois/Kentucky borders. This spot gets 2 min 41 s of totality in 2017 and 4 min 10 s in 2024 (see Fig. 6.16).

There really is only one country to go for the 2017 eclipse: the USA! The question is, as it crosses the entire country, from west to east, which state is the best? The umbral track first hits the US mainland at Lincoln City, Oregon where 2 min of totality will be visible on the centreline with the Sun already at an altitude of 40°. The state capital Salem goes under the shadow less than 2 min later. Moving inland may be necessary to avoid any coastal fog or cloud. The umbral track crosses Idaho, Wyoming, Nebraska, Missouri, the aforementioned Missouri/Illinois/Kentucky border region, Tennessee (clipping Nashville), the north Carolina/Georgia border regions and South Carolina. The track finally leaves the east coast just north of Charleston (which is right on the edge of the shadow) and peters out in the mid Atlantic. The greatest eclipse occurs at 36°58.0′N and 87°37.7′W close to Hopkinsville Kentucky and just east of the 2024 eclipse overlap point, with the Sun at an altitude of 64°. The full 2 min 40 s is potentially visible here, skies permitting.

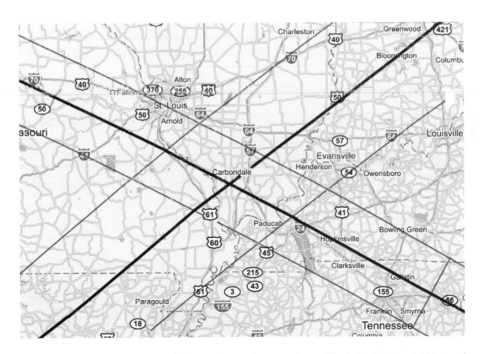

Fig. 6.16. The city of Carbondale, Illinois lies on the centre lines of both the 2017 and 2024 total solar eclipse tracks that cross the USA. Total eclipse tracks that crossover on land within such a short period are very rare, especially those that cross in a highly populated region or a city.

On the east coast, north of Charleston, totality drops to 2 min 35 s, with the Sun still at over 60° altitude in the early afternoon sky.

Where you observe this eclipse from will depend on whether you live in the USA and how far you live from the track, and if you are hell bent on travelling to a clear spot in the 24 h before the eclipse. If you are a visitor you may well want to spend a couple of weeks exploring the country, or simply pick a site nearest to other astronomical interests. If you go with an organised eclipse company specialising in eclipse trips for astronomers the itinerary will be decided for you. The obvious astronomical attractions include Meteor Crater, the Kitt Peak and Lowell observatories in Arizona, Mount Palomar and Mount Wilson Observatories in California, and the Very Large Array radio telescope in New Mexico. Arizona is obviously a key state for astronomy and is in the western USA. The best weather prospects for this eclipse are also in the west so one plan might be to visit the major astronomical attractions in Arizona and then head 1,000 km north to Idaho or Wyoming for the big event. Of course, much will depend on the weather prospects on the day and no doubt travel companies will hire thousands of coaches to head for wherever skies are clear. For the ultimate in flexibility a hire car is the best solution for small groups of eclipse chasers visiting the USA. During totality the Sun is flanked on either side by Mercury (12° to the east), and Mars (8° to the west). Venus will be sitting much further to the west (34°). The bright star Regulus in Leo (magnitude 1.4) will be just 1° to the east of the Sun during totality and may well be immersed in the corona!

2 July 2019 Total Eclipse

Statistics:

Saros: 58th of 82 from cycle 127.
S. Pacific eclipse ending in Argentina.
Maximum duration: 4 min 33 s in S. Pacific.
Shadow width at maximum: 200.5 km.
Moon larger by (Greatest): 4.592%.
Best destination: S. Pacific?

Two years after the total solar eclipse in the USA we have another long South Pacific Ocean event to look forward to. On first glance at the eclipse track you may well think you have seen this eclipse before! It is rather similar to the total eclipse of 11 July 2010. This eclipse track starts to the east of New Zealand, peaks 1,000 km north of Easter Island (the 2010 eclipse passed over the island) and ends in South America. However, unlike the 2010 event this eclipse ends much higher north on the South American continent and just about makes it to the eastern coastline. Figure 6.17 shows this and an incredible five additional tracks crossing southern South America between 2010 and 2027. Viewing this eclipse from Chile or Argentina is a realistic option as it is still at a decent altitude and a decent length as it crosses these countries. This eclipse track passes through the boundary between the French Polynesian and Pitcairn islands in the late morning phase of its journey, just prior to passing north through the tropic of Capricorn. The main Pitcairn island, a British territory, was very close to the track of the 8 April 2005

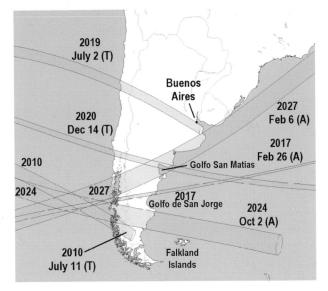

Fig. 6.17. Argentina would appear to be an ideal place to live if you are an eclipse chaser in the years 2010–2027. Three total and three annular tracks pass over Argentina and Chile, although the 11 July 2010 total eclipse only occurs at sunset in southwest Argentina, and is best viewed from Easter Island. Diagram by the author using *WinEclipse* by Heinz Scsibrany

hybrid eclipse. In fact Oeno island 120 km to the northwest was almost within the narrow totality centreline. Pitcairn island is only a few kilometres across and has less than 50 residents, who are descended from six of the mutineers from HMS Bounty in 1789. The island received some unwelcome publicity regarding alleged criminal activity around the time of the 2005 eclipse. Readers of Peter Hamilton's best selling "Reality Dysfunction" science fiction series may well recall the reincarnated mutineer character Fletcher Christian when they hear of HMS Bounty. Current predictions place Oeno island just inside the southern border of the umbral shadow this time. The 3.2 km diameter island is surrounded by a reef and a shallow coral lagoon. It is unspoilt and idyllic but there are no modern facilities and it is not geared for loads of visitors. Nevertheless it would be a beautiful place to view totality from. The umbra passes over Oeno at 18.26 U.T. with the Sun at an altitude of 32°, in the northeast. The duration of totality will be 3 min 25 s on the centerline, some 60 km northwest of Oeno. On the island over 2 min of totality should be experienced. An hour after passing Oeno the eclipse peaks at 108°57.0′W and 17°23.5′S in the southeast Pacific, with a full 4 min and 33 s available for those on a cruise ship. The Sun will be at an altitude of 50° from this point. Little more than an hour later, just before 20.40 U.T., the umbra hits the coastline of Chile, just north of La Serena and Coquimbo in the late afternoon. The Sun will be at an altitude of 13° above the Pacific ocean and totality will still last for 2 min and 35 s on the coast. Travelling at over 10,000 km/h the shadow takes less than 1 min to travel from the west Chile coast into Argentina. The cloud cover prospects for Chile and Argentina are similar to those of the nearby Pacific, i.e. 50–60%, but, of course, with the Sun being lower in the sky the chance of a cloud getting in the way is always greater. Only 3 min later the Sun sets in eastern Argentina and the umbra lifts off the Earth's surface into space. Just prior to sunset the shadow tracks slightly south of the Argentinian capital Buenos Aries before ending within a few kilometres of the Bahía Samborombóm bay area.

During totality the Sun will be in Gemini and Venus will lie 12° to its west, so the brightest planet will be impossible to detect from the Chile coast. Mars and Mercury will be elongated by 20 and 23°E, respectively.

14 December 2020 Total Eclipse

Statistics:

Saros: 23rd of 72 from cycle 142.
Crosses Pacific & Atlantic. Peaks in Argentina.
Maximum duration: 2 min 10 s in S. America.
Shadow width at maximum: 90.3 km.
Moon larger by (Greatest): 2.536%.
Best destination: Clear sky prospects fractionally better in Chile.

Almost 18 months after South America was host to an umbral track, it happens again in December 2020. But this time the track passes 1,000 km further south and occurs at midday (as before, consult Fig. 6.17). For eclipse chasing residents of Chile and Argentina 2019 and 2020 will be memorable years. This eclipse is the next one in Saros cycle 142 after the 4 December 2002 southern hemisphere eclipse which ended in the Australian outback. This eclipse starts from the north-east corner of the French Polynesian islands and heads southeast across the southeast Pacific Ocean before the central umbra hits the Chile coastline almost 100 km north of Valdivia, just south of the small town of Carahue. On the centreline 2 min 8 s of totality will be experienced. The track then passes south of the larger town of Temuco (near the north limit) before crossing the border with Argentina and the southern Andes. The maximum of this eclipse occurs at latitude 40°21.1′S and 67°54.4′W, with the Sun 73° above the north horizon and 2 min 10 s on offer. This position is roughly 150 km south of General Roca. The umbra then speeds towards the east Argentinian coast where the centreline meets the South Atlantic Ocean at the Golfo San Matías, slightly to the south of San Antonio Oeste. The northern edge of the track skirts the south facing bay here and even the centreline of the eclipse touches land again near Viedma, before finally continuing solely over water after 16.25 U.T. The path continues right across the South Atlantic Ocean and almost reaches the coast of Namibia in southern Africa. The track ends at sunset just 400 km west and slightly south of the Walvis Bay region, where the Sun will just have disappeared below the horizon. In an average December the prospects are slightly better, from a cloud cover perspective, on Argentina's east coast compared to the west coast. They are not too bad at the end of the track near Africa either, but here the Sun will be very low down unless your cruise ship is more than a 1,000 km west of that continent. Totality shrinks to only 25 s at sunset. Chile and Argentina look like the best bet for this eclipse. The weather prospects are only 50:50 at best along the track, although the prospects in Chile are fractionally better than elsewhere in an average December. Venus will be elongated by 24° to the east of the Sun for this eclipse. Mercury lies only 3° to the east so will be just outside the edge of the visible solar corona.

4 December 2021 Total Eclipse

Statistics:

Saros: 13th of 70 from cycle 152.
Crosses Antarctica.
Maximum duration: 1 min 54 s Antarctic coast.
Shadow width at maximum: 419.1 km.
Moon larger by (Greatest): 3.669%.
Best (only!) destination: Antarctica.

The previous eclipse in this Saros cycle, namely that of 23 November 2003, was surprisingly well attended despite occurring over Antarctica, in freezing conditions, and at a very low altitude. A few enterprising travel companies took almost 500 people onto the icy terrain or on plane flights under the shadow. A further 300 or so Antarctic scientists and engineers were under the shadow too. No doubt the same will happen for this eclipse. The reason that both these eclipses are in Antarctica is simply that they are very early (12th and 13th) in Saros cycle 152 and so both clip the bottom of the world. Whichever way you look at it the bottom of the world is Antarctica! However, consecutive eclipses in the same Saros cycle occur 8 hours later in the day and so the track shifts 120° to the west. In this case it means that the track favours the part of Antarctica due south of the tip of South America, Cape Horn and the Falkland Islands. The track can be seen on the global map (Fig. 6.1) at the start of the chapter and the Antarctica figure in Fig. 6.20. In 2003 the region near the Southern Ocean (below the Indian Ocean) was favoured. In fact, the track of this eclipse starts, at sunrise, only a few hundred kilometres east of the Falkland islands so this may well be a sane place to travel from to get to the eclipse, assuming a suitable cruise line takes up the challenge. The track starts at latitude 54°7.0′S, 49°15.8′W, but at this point the Sun is on the horizon. At its peak, on the Antarctic coast, 1 min and 54 s of totality is available with the Sun a healthy 17° above the horizon. Hopefully, being Antarctic summer, an ice breaker will not be necessary! With any eclipse to such a remote part of the world your options are limited by what the specialist travel companies will provide in 2021. At the current rate of global warming we have to hope that Antarctica has not melted by then!

20 April 2023 Hybrid Eclipse

Statistics:

Saros: 52nd of 80 from 129.
Clips W. Australia/Indonesia; ends in Pacific.
Maximum duration: 1 min 16 s of totality at East Timor.
Shadow width at maximum: 49.1 km.
Moon larger by (Greatest): 1.323%.
Best destination: Extreme W. Australia to East Timor.

This is the second hybrid eclipse covered in my brief summary of forthcoming totalities and, yet again, the eclipse is more total than annular. Indeed, the annular phase only occurs at the sunrise and sunset points. Starting way down in the Southern Ocean the umbra moves rapidly northeastwards and, most fortunately, just clips the extreme northwestern tip of Australia at the Cape Range National Park and Exmouth Gulf regions at roughly 22°30′S and 114°E. The track appears in Fig. 6.14 along with two other totals (already covered) and three annulars. At this Australia clipping point the umbral track is 41 km wide and totality 62 s long with the Sun already 54° above the NNE horizon. The umbra then speeds northeastwards across the Indian Ocean and Timor Sea and just brushes the easternmost end of East Timor. Fractionally before this point, at 9°36.0′S and 125°50.5′E the eclipse peaks with a track width of 49 km and duration of 76 s. The Sun will be 67° above the north horizon. The umbra then tracks across the Seram Sea before hitting Indonesian New Guinea, specifically, the region near the Teluk Berau strait just prior to the Doberai Peninsula (West Irian Jaya). The town of Babo is just north of the narrow umbral track. Shortly after this point the islands of Selat Yapen and Biak are grazed by the eclipse. The track then curves 600 km below the island of Palau and finally ends in the mid Pacific at sunset. The prospect of cloud along the track is lowest where it hits the northwest Australian coast and there should be a 60–70% of seeing the eclipse from there. Given the fact that Australia is a civilised country and there is only a loss of 14 s of totality this has to be the favourite land-based site from which to view this eclipse. Nowhere else along the track are the clear sky prospects any better than about 50% but the Australia to East Timor stretch offers a good 50:50 chance of seeing the eclipse. After East Timor cloud typically increases in April in this region.

8 April 2024 Total Eclipse

Statistics:

Saros: 30th of 71 from cycle 139.
Crosses Mexico, USA and Newfoundland.
Maximum duration: 4 min 28 s.
Shadow width at maximum: 197.5 km.
Moon larger by (Greatest): 5.655%.
Best destination: Mexico.

This is the second total solar eclipse across the USA in less than 7 years (see my earlier comments on the 21 August 2017 eclipse and Fig. 6.15) and, in terms of duration it is also substantially longer, 4 min 28 s maximum as opposed to 2 min 41 s. As previously mentioned and shown in Fig. 6.16, the two tracks cross close to Carbondale, Illinois. The 2017 track runs from the northwest to the southeast of the USA and it would be hard to imagine an eclipse track covering more US ground in more US states. The 2024 track runs from southwest to northeast, entering Mexico and then crossing through central Texas (including Dallas), the southeast tip of Oklahoma, central Arkansas, southeast Missouri, southern Illinois, southern Indiana and north-western Ohio. It then crosses the lakes Erie and Ontario where it travels along the US/Canadian border regions, with Toronto and

Montreal on the northern edge of the track. On the US side the umbra clips the northwestern edges of Pennsylvania, New York, Vermont (just) and Maine. Older residents of Maine and the Eastern US seaboard may remember the 7 March 1970 total solar eclipse three Saros periods in this cycle earlier, which took a similar, but more southerly track. We have already seen that this similarity with eclipses three Saros periods earlier is quite normal, as the 18 year, 11 day, 8 h time gap translates into 54 years and 34 days exactly, so the 8 h east–west longitude shift disappears, just leaving a typical shift of 1,000 km or so in latitude.

At the end of the North American track the 2024 umbra crosses New Brunswick before rapidly moving across the Gulf of St Lawrence and Newfoundland. The maximum of 4 min 28 s occurs at 25°16.9′N, 104°5.1′W in Durango Mexico, close to the towns of Gómez Palacio and Torrcon. Thus, for the USA, this is an afternoon eclipse. The weather prospects in Mexico, assuming the world's climate has not totally changed by 2024, are the best, with a better than 50% chance of clear skies. Conditions deteriorate statistically as the eclipse proceeds across the USA. Thus Texas offers the best US prospect and Pennsylvania to Maine the worst. Texas offers the best duration within the USA too, with 4 min 23 s on offer at Dallas, only 5 s short of the maximum. In the north Atlantic the track ends, at sunset, 1,000 km southwest of the southern coast of Ireland. Mercury will be 6° E and Venus 15° W of the Sun at totality. Jupiter should also be visible some 30° to the east.

12 August 2026 Total Eclipse

Statistics:

Saros: 48th of 72 from cycle 126.
Passes Greenland and the north Atlantic, ending in the Med.
Maximum duration: 2 min 18 s.
Shadow width at maximum: 293.7 km.
Moon larger by (Greatest): 3.867%.
Best destination: Northern Spain at sunset?

This eclipse is the next in the Saros cycle that we started this chapter with, i.e. the one 18 years 11 days and 8 h after the Russia/Mongolia/China eclipse on 1 August 2008. Yet again it is an eclipse that passes very close to British waters, although this time it does not touch any British islands like the Faeroes. Figure 6.18 shows the 2026 track and the ends of the 2028 totality and the 2027 annular. Just before the maximum point, as it passes down the east coast of Greenland the track is actually heading due south, as the umbra hits the Earth at a very high latitude, on the afternoon half of the globe. At the start the umbral track emerges from the terminator on the extreme northern coast of Siberia, passing rapidly over the arctic, just missing the north pole and then, as mentioned across Greenland. It just clips the west coast of Iceland before it plunges south passing within a few hundred kilometres of the extreme southwestern tip of Ireland. Just before the Sun sets, the eclipse, including its partial phases, can be seen in its entirety across the north coast of Spain. By the time the track enters the Mediterranean though, the final partial phases will be lost and totality itself is just denied, to the residents of Sardinia by mere minutes. From the north coast of Spain 1 min and 50 s of totality will be available with the

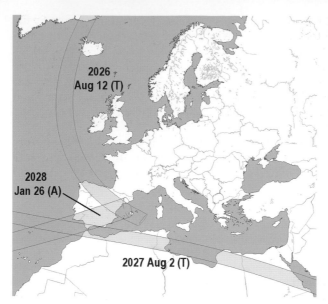

Fig. 6.18. From August 2026 to January 2028 two total eclipses and one annular eclipse can, cloud permitting, all be observed from Spain. The 2027 total eclipse and 2028 annular (just) can be seen from Gibraltar. The 2026 and 2028 events both end near the Balearic islands. The 2 August 2027 total eclipse is one of the really "big" (6 min 23 seconds) totalities of Saros 136 (number 38). Diagram by the author using *WinEclipse* by Heinz Scsibrany

Sun only 11° above the west horizon, i.e. *not* the sea horizon. Nevertheless, from a warmth and weather prospect, the north coast of Spain is probably the best bet, assuming you can find a spot with a good westerly view. Unusually, Jupiter is the closest planet to the Sun for this eclipse, although being 12° W of the Sun it will just be below the horizon from northern Spain, as will Mercury, which is 15° to the west. Venus will be at maximum elongation, 46° to the east of the Sun and blazing at magnitude −4.3. The holiday islands of Ibiza (just), Majorca and Minorca (just) fit within the wide shadow but from here the altitude at totality is only fractionally above the horizon. You need to be on a hill overlooking the western Med!

2 August 2027 Total Eclipse

Statistics:

Saros: 38th of 71 from cycle 136.
Follows (roughly) the North African coastline.
Maximum duration: 6 min 23 s.
Shadow width at maximum: 257.7 km.
Moon larger by (Greatest): 7.903%.
Best destination: Egypt.

We end this 19 year span of total solar eclipses with the next monster in the superb Saros cycle 136. The track of this eclipse is especially interesting as it passes right through the Straits of Gibraltar and literally hugs the north African coast. It even bends down to almost follow the Red Sea! (Refer to Fig. 6.18.) For European eclipse chasers in 2027 it will be especially attractive as it is not only of long duration and passing close to popular holiday destinations, but it also passes over some very sunny countries bordering the northern Med. For anyone who witnessed the 4 min

2 s of totality in Libya in 2006 it will bring back good (if gruelling) memories. Spain will experience its second total solar eclipse in the space of a year with this event, but this time it is the extreme southern tip and Gibraltar that is favoured. The umbra passes over the northern regions of Morocco, Algeria, Tunisia and Libya (West of the Gulf of Sirte), before plunging southeastwards across Egypt. Here, near the Red Sea Coast, just east of Luxor, the eclipse peaks at its spectacular 6 min and 23 s maximum, in a desert region renowned for its endless sunshine. Oh bliss, Oh joy! After the Red Sea the track crosses southwestern Saudi Arabia, skirting the border with North Yemen, before crossing the Gulf of Aden. It then clips the northeastern tip of Somalia before ending in the Indian Ocean. Brilliant Venus and Mercury will, respectively, be 3 and 10° W of the sun during the eclipse.

Annular Eclipses 2009–2028

I have heard the view expressed that an annular eclipse is the next best thing to a total solar eclipse. Well, if the Sun and Moon are so close in size that the eclipse is really a hybrid, or annular total, as some sources call it, maybe . . .. However, to me, an annular eclipse is a partial eclipse. It is like a total eclipse a few minutes before things get exciting. It doesn't get dark and the corona is not visible. "But, Venus appears" I hear you cry. Yes, but Venus is visible in broad daylight anyway, and I can see Venus at dawn or dusk for much of the year! Personally, I feel cheated with less than a minute of totality, so I guess I am a hard-core corona man. Still, there are quite a few people who chase around the world to see annulars too, so I might as well mention the forthcoming ones in passing. In general most eclipse chasers (except the diehard fanatics) will travel across the world to see totality, but will only travel to an annular if it occurs within a few thousand kilometres and from a civilised city that is not expensive to fly to. But despite what I have just said I would very much like to see an annular eclipse over the sea, at sunrise or sunset, when the low altitude dims the disc enough to view it direct, like watching out for the Green Flash. The "duration at greatest eclipse" statistic is worth a brief comment. The greatest eclipse point is when the axis of the Moon's shadow passes closest to the Earth's centre, essentially, near enough, the centre of the track. For many annulars, as for total eclipses, the duration is usually at its maximum at this point as the more central regions of the globe will rotate faster and the antumbra or umbra will be moving more slowly across the ground. However, unlike with a total eclipse, the track width will be smaller in the middle, as the Moon will look bigger in the central track and, if it got really big (i.e. the hybrid case) the antumbral track would change into an umbral track of infinitesimal width as annular became total. Things are a bit more complicated than this in extreme cases and where the shadow axis is at high latitudes or the eclipse a hybrid, but I hope you follow my reasoning. However, the "duration at greatest eclipse" for the thin ring annulars, approaching the hybrid case, is actually at a minimum, simply because the ring is so thin and therefore breaking the annulus is a lot easier when the Moon is slightly nearer, despite the slower antumbral speed across the ground in the centre of the track.

The longest possible duration of an annular eclipse (a Ring of Fire eclipse) is 5 min longer than for the longest possible totality, namely 12 min and 30 s (compared to 7 min and 31 s). On 14 December 1955 a 12 min 9 s Ring of Fire annular occurred

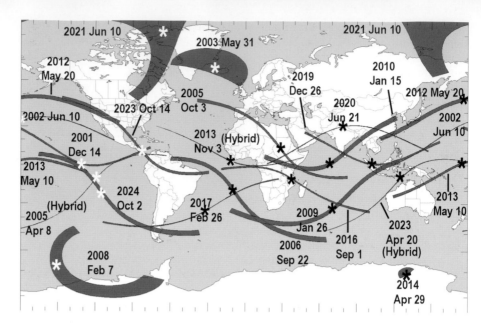

Fig. 6.19. The tracks of all annular and hybrid (marked as such) eclipses from 2001 to 2025. *Asterisks* mark the point of maximum eclipse. A modified version of a diagram by Fred Espenak NASA/GSFC. Diagram by the author using *WinEclipse* by Heinz Scsibrany.

Table 6.2. Annular eclipses from 2009 to 2028, arranged in duration order

2010	2028	2009	2027	2024	2013	2012	2023	2021	2019	2016	2026	2017	2020	2014
11 min 8 s	10 min 27 s	7 min 53 s	7 min 51 s	7 min 25 s	6 min 4 s	5 min 46 s	5 min 17 s	3 min 51 s	3 min 39 s	3 min 5 s	2 min 20 s	44 s	38 s	Zero

but the longest annular in this section is a minute shy of that one. Figure 6.19 shows the tracks of all the annular eclipses (and three hybrids) between 2001 and 2025 .

26 January 2009

Statistics:

Saros: 51st of 71 from cycle 131.
Crosses Indian Ocean ending near Borneo.
Duration at greatest eclipse: 7 min 53 s.
Annular track width at maximum: 280.2 km.
Moon smaller by (Greatest): 7.175%.
Best destination: Southern Indian Ocean.

Essentially, with an ocean-crossing track, far from the USA and Europe, this annular is unlikely to attract too many travellers. However, if you happen to be on a ship in the Celebes Sea west of Borneo on 26 January 2009, and if skies are clear (very unlikely) the sunset view of this annular could be very memorable indeed. Refer to Fig. 6.14 for the track.

15 January 2010

Statistics:

Saros: 23rd of 70 from 141.
Starts in Africa; ends in China.
Duration at greatest eclipse: 11 min 8 s.
Annular track width at maximum: 333.2 km.
Moon smaller by (Greatest): 8.098%.
Best destination: Maldives or Burma.

This is a real "Ring of Fire" eclipse, i.e. one in which the Moon appears much smaller than the Sun because the Moon is close to apogee and the Sun is close to perihelion. On the negative side it is about as far from a total eclipse as you can get, with a considerable amount of sunlight still reaching the Earth and, although the illumination will be weird, the overall brightness levels will be as high as on a cloudy day. The start of the track in Africa is covered in Fig. 6.11. Remarkably, once again, eastern China is along the final part of the track. That region also experiences total eclipses on 1 August 2008 (at sunset) and 22 July 2009 (refer back to Figs. 6.4 and 6.6) as well as an annular eclipse that crosses Hong Kong on 20/21 May as described in the next section. The town of Zhongxian at longitude 108°E and 30°20′N is near the centre line of this eclipse and near the total eclipse of 22 July 6 months earlier! Residents of Australia's north Queensland, living between Rutland Plains and Old Strathgordon have the same experience 3 years later in 2012/2013.

20 May 2012

Statistics:

Saros: 58th of 73 from cycle 128.
Starts in China, crosses Japan, and ends in USA.
Duration at greatest eclipse: 5 min 46 s.
Annular track width at maximum: 236.9 km.
Moon smaller by (Greatest): 5.611%.
Best destination: Nevada/Utah/Arizona.

China's fourth major eclipse in as many years starts in that country at sunrise and includes Hong Kong within its southern boundary, where 3 min 27 s of annularity is predicted. The eclipse also crosses the whole of the southeastern side of Japan, including Tokyo, where the duration is 5 min and 4 s. The final stages are in the USA in the late afternoon (covered in Fig. 6.15). The track enters southern Oregon and northern California, passing through Nevada (and the city of Reno), Utah and Arizona, as well as New Mexico (including Albuquerque). The eclipse ends, at sunset, in Texas, just after passing over Brownfield. Although this is only an annular eclipse it may well be a good excuse to visit those US landmarks and astronomical attractions that you have not yet been to. The track of annularity in the unreal evening illumination of that day crosses the Grand Canyon, just skirts north of Flagstaff and the Barringer Meteor Crater near Winslow. Earlier on along the track the centreline passes over Lassen Peak in Shasta County California, the largest single lava dome volcano in the world (height 3,189 m) and the only other volcano in the Cascade mountain range to have erupted in the twentieth century (1914–1917), apart from Mt St Helens. So, all in all, this is a good annular eclipse to go to even if, like me, you regard annular eclipses as nothing more than deep partials.

10 May 2013

Statistics:

Saros: 31st of 70 from cycle 138.
Starts in W. Australia; ends in E. Pacific.
Duration at greatest eclipse: 6 min 3 s.
Annular track width at maximum: 172.6 km.
Moon smaller by (Greatest): 4.559%.
Best destination: W. Australian Desert.

As well as Eastern China bagging more than its fair share of eclipses in the previous 4 years, northern Queensland decides to get in on the act. As we saw in the previous section, only 6 months after the region between Rutland Plains and Old Strathgordon was host to a total eclipse at sunrise, the same region experiences an annular in the early morning (Fig. 6.10). However, this time Western Australia and the Northern Territory region get a good view too. After leaving Australia the zone of annularity passes over the extreme eastern tip of Papua New Guinea, a small island named "Woodlark Island" and the Solomon islands before ending in the East Pacific.

29 April 2014

Statistics:

Saros: 21st of 75 from cycle 148.
Confined to a point in Antarctica!
Duration at greatest eclipse: 0 s!
Annular track width at maximum: 0.0 km.
Moon smaller by (Greatest): 1.447%.
Only (!) destination: Antarctica.

This has to be the weirdest eclipse in this book, and, unless you are an Antarctic researcher, definitely *not* worth the effort of travelling to (see Fig. 6.20). Essentially, this eclipse represents the absolute on-the-limit condition when a partial eclipse just classifies as an annular (or, if the Moon were bigger, a total eclipse). The zone of

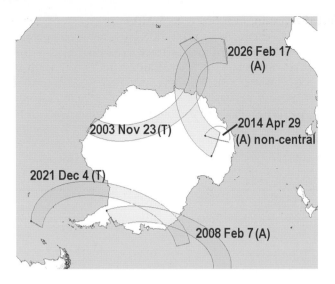

Fig. 6.20. Two total and three annular eclipses cross Antarctica between 2003 and 2026. The 2003 totality was well attended by dedicated umbraphiles, so maybe the 2021 totality will attract the hard core eclipse chaser too. The 2008 annular event was imminent as this book reached publication and, of course, the 2026 event is the next eclipse in that Saros. As described in the text the 2014 event is just plain weird, a glancing, non-central annular eclipse of 0 s duration! The tips of South America (*lower left*), Southern Africa (*upper left*), Australia (*upper right*) and New Zealand (*lower right*) are included for orientation purposes. Diagram by the author using WinEclipse by Heinz Scsibrany

annularity *just* hits Wilkes land in the Antarctic. Indeed, none of the other 20 eclipses in this Saros achieved even this; they were all partials from the Antarctic region. Technically, this eclipse is a rarity from a geometric perspective too. It is the only one in this book where the shadow axis misses the Earth's surface and yet, it still classifies as an annular. From the observers viewpoint (and, frankly, I think the observers will be penguins!) this technicality means that the brief infinitesimal period of annularity will not show the Moon concentric within the "ring of fire". For the briefest moment a lopsided ring of light will encircle the Moon, which will be on the horizon, but blink and you will miss it. Only eight such non-central eclipses occur in the 1900–2100 two hundred year period. Oh, and you will need to be at exactly 70°41.8′S 131°9.5′E. Being in a helicopter hovering above that point will help! So, are you crazy enough to go to this one? Someone will be! My advice: Get a life.

1 September 2016

Statistics:

Saros: 39th of 71 from cycle 135.
Crosses southern Africa & Madagascar.
Duration at greatest eclipse: 3 min 5 s.
Annular track width at maximum: 99.7 km.
Moon smaller by (Greatest): 2.637%.
Best destination: Madagascar.

This run-of-the-mill annular eclipse tracks across Africa and northern Madagascar (see Fig. 6.11), crossing Gabon, Congo, D.R. Congo, Tanzania, Malawi and Madagascar. It ends in the south Indian Ocean not far from the coast of SW Australia. There is a 70% chance of clear skies on the west coast of Madagascar.

26 February 2017

Statistics:

Saros: 29th of 71 from Saros 140.
Crosses Chile, Argentina, S. Atlantic; ends in D.R. Congo.
Duration at greatest eclipse: 44 s.
Annular track width at maximum: 30.6 km.
Moon smaller by (Greatest): 0.778%.
Best destination: S. Atlantic for thinnest ring.

Okay, forget what I said about annulars! This one and the one in 2020 look fairly interesting. This does not classify as a hybrid eclipse, because the Moon is smaller than the Sun all along the track. However, it is not much smaller, especially at mid-eclipse, where the ground track of annularity is only 30 km wide. Members of the International Occultation Timing Association (IOTA) are keen on travelling to the edges of totality tracks where video timings can help in analyzing data about the precise size of the Sun (see Chap. 10). For this eclipse, the track edges are virtually on the centreline. At the point of greatest eclipse, in the South Atlantic (34°41.6′S, 31°8.4′W) where the duration of the annular phase is actually at its least, the Moon is only 0.778% smaller than the Sun. This is equivalent to an additional radius of less than 15 km at the distance of the Moon. Thus, with, appropriate filters, the Sun will be the merest sliver of a ring around the lunar edge, with the tallest peaks threatening to break the ring up. NASA's Fred Espenak, in his predictions, uses a value for the lunar radius that ensures all total solar eclipses are total, i.e. it accounts for the deepest lunar valleys. Thus the actual rugged mountain-peaked Moon may be fractionally bigger in practice. The figure of 44 s is thus a bit mean-ingless as any unaccounted for towering peaks on the limb could technically break the annulus early, reducing the actual period of true annularity. Unfortunately, even with the annulus this thin, the eclipse will be far too dangerous to look at without filtration. The annular crosses Chile and Argentina (Fig. 6.17), the Atlantic, Angola and D.R. Congo (Fig. 6.11). Chile and Argentina do very well from 2010 to 2020 with three total eclipses and this annular crossing (or ending) in these countries. For the six tracks across this region, from 2010 to 2027, consult Fig. 6.17.

26 December 2019

Statistics:

Saros: 46th of 71 from Saros 132.
Starts in Saudi & crosses India, Sumatra, Malaysia & Borneo.
Duration at greatest eclipse: 3 min 39 s.

Annular track width at maximum: 117.7 km
Moon smaller by (Greatest): 2.987%.
Best destination: Saudi Arabia at Sunrise?

The track of this northern hemisphere annular starts near the Saudi Arabia/Qatar border and crosses the United Arab Emirates and Oman, the Arabian Sea, southern India and northern Sri Lanka. After crossing the Indian Ocean, it arrives at Indonesia (Fig. 6.14) and the northern limit just skirts Singapore before the track follows the Sarawak/Borneo border and just misses the southern Philippines island of Mindanao. After passing the island of Palau the track ends in the sea on the border of the northern Mariana and Marshall islands. A pleasant location to watch the annular Sun set in the Pacific.

21 June 2020

Statistics:

Saros: 36th of 70 from Saros 137.
Crosses Africa, Arabian Peninsula, N. India & China.
Duration at greatest eclipse: 38 s.
Annular track width at maximum: 21.2 km.
Moon smaller by (Greatest): 0.600%.
Best destination: India/China border for thinnest ring.

This is another ultra-thin annulus eclipse and a lunar radius only 10 km greater would turn this into an annular total. Yet again the precise duration of annularity in the middle of the track may be modified by the odd massive lunar peak on the limb breaking the annulus. For once the track passes mainly overland; indeed it seems to curve to avoid the sea! The track covers a huge number of countries in its passage and if only the Moon were fractionally closer it would permit huge numbers of people to see the corona. Sadly, the thin annulus will still be blindingly intense. The track starts in Africa on the Congo/D.R. Congo border (Fig. 6.11). It then proceeds through Sudan, Ethiopia, Yemen, southern Saudi Arabia, Oman, Pakistan, India and China (yet again!) and Taiwan. The maximum point, thinnest ring, narrowest track, and, shortest duration of annularity for this thin ring event, occurs in the Himalayas at 30°31.0′N and 79°43.2′E in northern India, literally on the border with Tibet and just west of the Nepal border. Unfortunately, even the height of the Himalayas will not be sufficient to raise the observer enough to turn this into a total eclipse!

10 June 2021

Statistics:

Saros: 23rd of 80 from Saros 147.
Starts in Canada, crosses north pole, ends in E. Siberia.
Duration at greatest eclipse: 3 min 51 s.

Annular track width at maximum: 526.6 km.
Moon smaller by (Greatest): 5.652%.
Best destination: Canada for accessibility.

Somehow I cannot see many people travelling to the frozen north to witness this eclipse. The post-sunrise part of the track in Canada is the only sane place to view it from, namely from Ontario, before the antumbra sweeps over the eastern Hudson Bay and James Bay (see Fig. 6.15).

14 October 2023

Statistics:

Saros: 44th of 71 from cycle 134.
Crosses USA, leaving at Texas. Ends off Brazilian coast.
Duration at greatest eclipse: 5 min 17 s.
Annular track width at maximum: 187.4 km.
Moon smaller by (Greatest): 4.796%.
Best destination: Utah/New Mexico.

As mentioned in the previous section, 2017–2024 is a golden period in the USA for eclipse, with two totals and this annular. In addition, it is less than 12 years since the USA annular of 20 May 2012. This annular traverses the states of Oregon, Nevada and Utah, skims the borders of Arizona and Colorado, crosses New Mexico and, once again, passes straight through Texas (again, see Fig. 6.15). The track leaves the Texas coastline near that lucky spot on the border with Mexico (see my comments regarding the 21 August 2017 total eclipse). After the USA the track passes over Yucatan, Honduras, Nicaragua, Panama, Colombia and Brazil before ending in the South Atlantic. Should be a popular one.

2 October 2024

Statistics:

Saros: 17th of 70 from cycle 144.
Crosses S.E. Pacific, ending off Argentina.
Duration at greatest eclipse: 7 min 25 s.
Annular track width at maximum: 266.6 km.
Moon smaller by (Greatest): 6.741%
Best destination: Sunset off Argentina?

Not a lot you can say about this annular. It is essentially over the Pacific Ocean, except for the very end phase where it passes over, yes, you've guessed it, Chile and Argentina! The very last part of the track just passes to the north of the Falklands and then sets in the South Atlantic Ocean (refer to Fig. 6.17).

17 February 2026

Statistics:

Saros: 61st of 71 from cycle 121.
Crosses Antarctic below Australia.
Duration at greatest eclipse: 2 min 20 s.
Annular track width at maximum: 618.2 km.
Moon smaller by (Greatest): 3.703%.
Best destination: Antarctic.

Another eclipse for the penguins I feel! Antarctic eclipses are, inevitably, near the start or finish point of a Saros cycle and this one is the 61st of cycle 71, in other words, near the end. See Fig. 6.20.

6 February 2027

Statistics:

Saros: 52nd of 71 from cycle 131.
Crosses South American coast, the mid Atlantic and ends (just) in Africa.
Duration at greatest eclipse: 7 min 51 s.
Annular track width at maximum: 281.6 km.
Moon smaller by (Greatest): 7.189%.
Best destination: South America.

What is it about the southern part of Chile and Argentina that attracts so many eclipse tracks between 2010 and 2027. Well, it is just pot luck. Nothing more, nothing less. Still, it seems a bit spooky to me. This annular passes over Chile and Argentina in the morning phase and then just hugs the coastline of South America as it heads northeast. It clips the coast of Uruguay before heading across the Atlantic and ending, at sunset, on the west and south Nigerian coasts. See Fig. 6.17 yet again!

26 January 2028

Statistics:

Saros: 24th of 70 from cycle 141.
Crosses northern South America and ends in Spain.
Duration at greatest eclipse: 10 min 27 s.
Annular track width at maximum: 323.1 km.
Moon smaller by (Greatest): 7.921%.
Best destination: Balearics for the sunset?

This is another "Ring of Fire" annular, with the Sun much larger than the Moon. The only comparable annular in my selection is the 2010 event. The track crosses Ecuador, Peru, Brazil and French Guina (where it peaks) and then heads across the Atlantic. On arrival in southern Europe it crosses southern Portugal and south and central Spain with the eclipse ending, at sunset, in the extreme western Mediterranean, near the Balearic Islands (refer to Fig. 6.18). A sunset viewing from this point would be a memorable one and not dissimilar to the 12 August 2026 sunset, except that annulars can only be safely viewed directly when rising or setting in the zero altitude horizon of an ocean. In the space of 6 months southern Spain and Gibraltar (right on this annular's southern limit) get a very long total eclipse followed by a very long annular.

Part II

Observing and Travelling to Total Solar Eclipses

Chapter 7

Safety First

No book about observing the Sun, whether eclipsed or not, would be fit for publication without mentioning the dangers of solar observing. These dangers were hammered home to me, many years ago, when the English observer J. Hedley Robinson damaged his eyesight while observing the Sun. I would not have believed such an experienced observer could have come to harm in this way, but he did. He was using an experimental filter he had devised to allow comfortable solar observing. However, he had tragically underestimated the amount of infra-red radiation the filter was letting through. The Sun looked pleasantly dimmed visually, but your eyes cannot see in the infra-red. Hedley became aware that his eye suddenly felt uncomfortably hot and instinct made him pull away. His vision in that eye was never the same. The eye is not equipped with pain sensors and so even when damage is being done to the retina it is not immediately apparent.

Most people, astronomers or non-astronomers, know that the Sun is dangerous. Even if the eye has no pain sensors human beings turn away from such a dazzling object. With the exception of young children, 99.9% of people surely know just how damaging it would be to look directly at the Sun, especially through a telescope or binoculars. So how do accidents actually occur? In virtually all cases, serious eye damage results when observers have deliberately overruled either their instincts or their intelligence, because they have been under a misapprehension. It is a case of a little knowledge being a very dangerous thing, or over-confidence in your understanding creeping in. The vast majority of eclipse-related eye injuries over the years have been sustained by people just staring at the Sun with the naked eye. There is a subconscious belief amongst some people that goes along the following lines: The Sun is always in the sky on a sunny day, we see it every day; therefore it is only dangerous through a telescope. *Wrong*! This misconception does not take into account the fact that while the Sun may fleetingly cross our field of vision on a sunny day, we do not stare at it. It is fuelled by the fact that when the Sun is partially eclipsed and low down, a few people without appropriate filters have decided to screw their eyes up against the dazzle and actually glimpsed the partial phase. Okay, there is an after-image, but it has not put them off! The high resolution part of the retina, called the fovea, is the part we all use for reading and high-resolution work and, of course, that tiny section is the area we use to stare at something. Once damaged by thermal lesions the retina can never be repaired. Hold a magnifying glass to focus the Sun on a piece of paper. After a few seconds the paper burns. On a lesser scale that is what you do when you stare at the Sun. Again, the lack of pain sensors in the retina combined with a small amount of knowledge is a lethal combination, as are the sheer numbers of observers involved. Well publicised eclipses like that of 11 August 1999 that cross through many highly populated countries will attract millions of observers. Statistically, dozens of people permanently damage their eyesight at such events; a small percentage yes, but devastating for the individuals concerned. The Sun is still dangerous even at a few degrees altitude. Only at precise sunrise and sunset is the light and heat attenuated enough for it to be safe to look at, when the opacity of the atmosphere dims the ball to the standard dull red sunrise/sunset colour we are all familiar with. Even then, considerable common sense should be exercised and you should *never* look at the Sun through a telescope without appropriate filters. When a solar eclipse is in progress you should be using proven equipment and proven filters that you have used many times in the past. In the heat of planning for an eclipse trip it sometimes transpires that you may lash something together that is not properly tested. You cannot afford to do this with solar work.

Another often unappreciated fact is that the single lens reflex viewfinder on a modern SLR or digital SLR (DSLR) camera is looking through the lens. In this digital age it can sometimes be imagined (literally in the heat of the moment) that all viewfinders are LCDs. They are *not*. Light passes into your eye straight from the sky or the telescope when you look through an SLR/DSLR optical viewfinder. Two dangerous periods are unique to total solar eclipses. They are second contact and third contact. They are periods when, within the space of a few seconds, the Sun goes from being lethal to perfectly safe and from perfectly safe back to lethal again. We are always encouraged to observe the diamond ring effect, especially at third contact, and Baily's beads too. However, being technically accurate, these *do* involve looking directly at the solar surface even if it is a tiny fraction of the surface seen through arcsecond gaps which are lunar valleys. I have lost count of the number of times I have heard the question asked "When is it safe to look directly at the Sun as second contact approaches?" or "How long can you stare at the final diamond ring before it is dangerous?" These are not questions that have simple answers like "4.73 s". It is simply common sense that will guide you. If the second contact (shrinking) or third contact (growing) Sun is too dazzling to look at, then do not look at it. As the whoops and cries go up as the light levels start to plummet, 30 s before second contact, you will instinctively know, by holding up your safe solar viewer, when that last sliver of a crescent is becoming a solitary point. The crescent, viewed through your filtered glasses, suddenly rapidly disappears and then you look. If it is too dazzling, you look away. The merest glimpse of a hairline thin crescent with the naked eye is not going to blind you. It is staring at the dazzling Sun or using inappropriate filters for many seconds that causes the damage.

Naked Eye Filters

In the build up to any solar eclipse you will often see adverts for solar filters and solar filter glasses. They come in a variety of shapes and sizes and, in recent years, equipping cuddly toys with them has also become quite fashionable. For proof, refer to Fig. 7.1. The spectacle versions are, invariably, cardboard frame glasses with silvery Mylar films over each eye section, in other words, ultra-dense cardboard sunglasses. For a typical supplier see Rainbow Symphony's website at http://www.rainbowsymphony.com/soleclipse.html. These types of inexpensive

Fig. 7.1. Man-mountain Bob Priest (plus furry filtered friend), photographed by a person of normal height, observing the partial phases of the total solar eclipse of 29 March 2006 with a safe British Astronomical Association solar filter. Image: Maria and Bob Priest

glasses are occasionally advertised as neutral (i.e. not coloured) density 5 (ND5), but when tested they are often safer, roughly ND6, so they typically attenuate the light by roughly one million times or 10^6. Do *not* assume any glasses are safe, especially if you purchase them abroad, in a third world country. In fact, *never* buy such glasses in a third world country. The basic safety test for such glasses is to look through the curtains of a darkened room at the outside, sunny, daylight world. The most you should be able to see is the merest ghostly hint of any detail and, preferably, not even that. Now look through the glasses directly at a really bright electric bulb, say, 100 W, from a foot or so away. You should just be able to see the electric filament glowing faintly. If the glasses pass this test, look through them at the Sun on a clear day, when it is high up. The Sun should be comfortable to look at and should never dazzle. If it dazzles, look away and bin the glasses. Also, examine the shiny Mylar surfaces of the glasses for pinholes; tiny areas of damage where the coating has been scratched off. If the coating is damaged do *not* use the glasses. If the glasses pass these tests then you can feel confident to use them *for a few seconds* at a time to check on the progress of the eclipse. But I would never recommend looking for any longer with such cheap devices. Never stare at the Sun through any kind of Mylar film filter. Eyes cannot be replaced. In the UK, a large number of solar filter glasses were manufactured for the 1999 TSE.

The good ones had a CE mark (Conformite Europeene) on the side to show the product had met European Union safety and standards. While this does not guarantee that they have avoided being scratched before you acquired them, it is reassuring if the glasses have been tested to an accepted safety standard. Incidentally, many people know that if you take two pairs of polaroid sunglasses and hold the lenses at 90° to each other, the darkening factor increases dramatically. Do *not*, under any circumstances, be tempted to use this lethal combination as a solar filter.

In October 2005, my eclipse-chasing friend Nigel Evans went to the annular eclipse which crossed Portugal, Spain and North Africa. He decided to observe the event from Tunisia. There he noticed the local traders had decided to exploit the opportunity and market their own solar viewer sunglasses. Nigel purchased a pair as a memento. Out of interest (he has plenty of solar filters) he decided to look through them to see how suitable they were. He was staggered. They were barely dimmer than a normal pair of sunglasses. He showed them to me shortly after he returned. I would estimate that allotting them an ND rating of 1 would be generous, let alone 5 or 6! Anyone staring at the partial phases with such glasses would be ordering a guide dog online the next day (using a Braille keyboard). Nowhere on the glasses did it say "Manufactured by the Burning Retina Optics Company", but that is how they should have been branded. I shudder to think how many local people might have suffered serious eye damage from such glasses, especially in a country where few people would know any better. Needless to say, being Tunisia, the skies were clear on the day. It would have been better, perhaps, if it had been cloudy.

I have never been too happy with cardboard frame Mylar sunglasses in any form, for various reasons. Firstly they come with one eye spacing. As someone with very close together eyes I notice these things. The famous optician Horace Dall (1901–1986) told me he had never met any adult with eyes as close together as mine and therefore I was "almost certainly a psychopath!" So, standard spacing glasses are as useful to me as the proverbial rubber pick axe, inflatable dartboard or chocolate teapot. I am not the only person with this problem. There are millions of psychopaths like me out there. In addition some people have wide apart eyes and, of course, all children have eyes that are closer together than adult eyes. If one eye is not totally covered there is danger present. Another thing about these cardboard glasses is that they are incredibly light. In my experience many eclipse sites are quite breezy, unsheltered places. Lightweight filters and cardboard glasses can easily blow off which is, at best, a hassle and, at worst, dangerous. However, there are a variety of naked eye solar filter designs to choose from. Some of the variants are shown in Fig. 7.2. Personally, I prefer the big black slab of ND6 polymer filters that you often see at solar eclipses. Being black they are not annoyingly reflective like the Mylar glasses. The glasses have to literally be squashed flat on your head to prevent you seeing a reflection of yourself. No such problem occurs with the black plastic filters. In addition, with a big 20×5 cm visor of plastic available both of your eyes are covered, whatever your eye spacing. Also, the polymer is thick and even when scratched they will still be safe.

Genuine welding goggles can also be used for brief *naked eye* eclipse viewing. The welding filter shade 14 is roughly equivalent to slightly brighter than neutral density 6, while shade 12 is slightly brighter than neutral density 5 and should be used with a lot more caution.

Fig. 7.2. A variety of safe naked eye solar viewers, the CE mark (Conformite Europeene) shows the product has met European Union health, safety and environmental standards. Two different types of filter material are shown, multi-layer aluminised Mylar (silver colour) and polymer filters (black). The single slit filters are preferable as they do not assume one size of eye spacing/observer's face. Image by the author

Projecting the Sun

Of course, the safest method of observing the partial phases is by projecting the image, using a small refracting telescope, onto white cardboard. I have been using a childhood 30 mm aperture refractor for this purpose for decades. Even smaller aperture binoculars can be employed. A cardboard screen around the telescope tube acts as a good sunshield so that the card on which the solar image is actually projected is kept in the shade. Otherwise the solar disc will appear ghostly. Direct sunlight will swamp the solar image coming through the telescope and displayed on the screen. The cardboard shield can be used for another purpose if it is near the eyepiece end of the refractor. When positioning the refractor at the Sun the shadow of the tube falling on the shield can tell you when you are close to the Sun. When perfectly aligned, the shadow will disappear into a concentric ring. Since the time of the European 11 August 1999 total solar eclipse a variety of commercial solar projection gadgets have appeared. Perhaps the best is the *Solarscope*, shown in Fig. 7.3, which comes in flat-pack "kit" form and is great for partial phase viewing by small groups of people.

Eye safety can sometimes swing the other way and lead to total paranoia about enjoying the total phases. I well remember being on the beach at Knip bay, Curacao, for the 1998 Caribbean eclipse. Some children had obviously been well versed in the dangers of the Sun: They were still holding the solar filter glasses up even when totality occurred. Fortunately, someone spotted them and they were able to enjoy totality. It is perfectly safe to observe the totally eclipsed Sun with the naked eye, binoculars, or an unfiltered telescope when it is completely hidden behind the Sun, but you *must* be ever-vigilant for the third contact diamond ring returning. Through a telescope, even a split-second of a sliver of sunlight can permanently scar the retina.

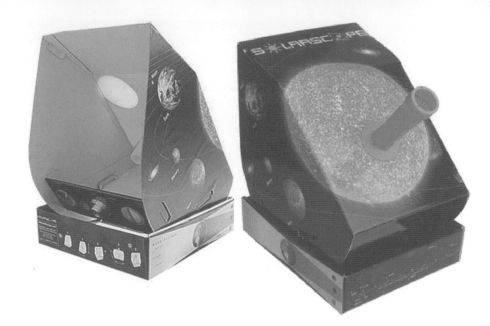

Fig. 7.3. The solarscope is an ultra-safe and educational way of viewing the Sun and is suitable for use by small groups of people. A 40 mm aperture lens focuses the image of the Sun, via a convex mirror, onto the inside of a card box at an effective focal length of roughly 9 m (f/230) resulting in an 80 mm diameter solar image. The whole unit comes in flatpack kit form. Image: Solarscope

Telescope Filters

A substantial proportion of the Sun's light and heat must be rejected by filtering. So, which solar filters are the best to use? Choosing a solar filter for a telescope is a serious issue. I can remember a time, in the 1960s and 1970s, when almost every cheap and nasty shopping mall refractor came supplied with a lethal solar filter. These horrors screwed into the eyepiece barrel, where all the heat concentrates, and, not surprisingly, they were liable to crack due to the heat falling on them. Thankfully, these nightmare filters appear to have become extinct and all modern solar filters are full aperture ones. By full aperture I mean that they fit at the very front of the telescope. With a refractor this means over the dew cap or the objective lens. Many telescope manufacturers supply full aperture solar filters that are specifically designed to tightly fit, without the risk of a gap, over their own instruments. The larger solar filter suppliers also sell filters designed to fit the most purchased telescopes, such as Schmidt-Cassegrains (SCTs) by Celestron or Meade, popular refractors, and the ETX model Maksutovs. A filter that fits tightly across the telescope aperture is essential. The so-called "off-axis" solar filters are also available for large aperture telescopes like SCTs. With these filters most of the aperture is blocked completely with a solid, opaque material, but a short, off-axis aperture (between the secondary mirror holder and the aperture rim) is filtered, thus turning a large Schmidt-Cassegrain, Maksutov, or even a Newtonian into a smaller aperture solar telescope. However, for those with deep pockets even 35-cm aperture SCTs can be fitted with full aperture filters. Take a look at Fig. 9.4.

For digital photography choosing a solar filter that lets through a bit more light than a safe visual filter can be a distinct advantage, but such filters should *never* be used visually. As we have seen, solar filters are often given an ND rating. This stands for neutral density and is graded logarithmically. I will repeat my explanation for safety's sake: An ND 1 filter attenuates by a factor of 10; ND 2 = 100 × attenuation; ND 3 = 1000 × attenuation. The standard safe visual filters are approximately ND5 or ND6; in other words, only one part in a hundred thousand or one part in a million of the light gets through to the observer's eye. Put another way 99.9999% of the light is blocked with an ND6 filter. The big six names in solar filter supplies are Baader, Thousand Oaks, Kendrick, Roger W. Tuthill, Orion Telescopes & Binoculars and Celestron. Thousand Oaks and Kendrick will sell you less dense filters, yielding a ten times brighter image for photography, CCD or even webcam use (for high resolution work). Obviously these should never be used visually unless specifically guaranteed for that purpose by the manufacturer. In fact, Celestron and Kendrick use Baader's AstroSolar film for the basis of their full aperture filters. Essentially, there are two types of safe solar filters in use by amateurs. The first type consists of an optically flat glass window. That type has a thin layer of aluminium deposited onto the glass (like the aluminium on your telescope mirror) but just thin enough to let, at most, 0.001% of the light through (see Fig. 7.4). The second type of telescope filter is an ultra-thin aluminised Mylar film, very similar to the material used in the naked eye glasses. This is cheap, because it is far easier to make aluminised Mylar than to mass produce optically flat glass discs and then aluminise them. It might be assumed that the glass filters are better, especially when you look at the rippling, crinkly, Mylar filters. However, tests do not bear this out. The damage that will be done to the incoming light by an ultra-thin sheet of Mylar seems to be no worse than that inflicted by a much thicker, but flat, metal-on-glass filter, especially when compared to the damage wreaked by the Earth's daytime atmosphere. From a safety viewpoint though, Mylar film is easily creased and damaged and should always be carefully inspected

Fig. 7.4. Full aperture "metal-on-glass" (Inconel) solar filters are the safest filters to use. Here two are fitted to a Russian 1,000 mm mirror lens and 90 mm aperture Maksutov, owned by Nigel Evans. Image by the author

before use. Of course, if you only ever observe the Sun with the webcam, safety may be less of a concern as a burning webcam can be replaced; a damaged retina cannot. Personally speaking, I only feel safe with a proper metal-on-glass type filter. With ND4 solar filters the extra bright image can result in webcam exposures as short as 1/1,000 of a second being achieved: ideal for freezing the seeing and capturing the finest hi-res solar details for non-eclipse imaging. Far too bright for safe visual use though. As well as revealing sunspot detail, a white light solar filter will record faculae (brighter, relatively white regions) and the solar limb darkening.

If you have the slightest doubt about any filter, *never* use it. Your retina cannot be repaired. Unless you have purchased your filters and equipment from a reputable astronomical supplier with a CE mark or a guarantee of transmission, you are taking a risk. The bright light bulb test described earlier should always be used as a double check. For more technical information, see http://www.mreclipse.com/Special/filters.html.

Chapter 8

Eclipse Trips – The Real Experience

It might be thought that total solar eclipse trips are simply astronomical expeditions, i.e. hundreds (and thousands) of like-minded amateur astronomers heading to see the greatest show on Earth for a few brief minutes. However, after you have been on a few eclipse holidays you realise that they are much more than that. Totality is a painfully short experience. Even when you have been to quite a few eclipses you are still shocked when totality ends. Spectacular though the event is, few people (except the diehard umbraphiles) plan for years and spend thousands of dollars just for a buzz, lasting a few minutes. Unless a total solar eclipse just happens to occur very close to your part of the world, travelling to one will probably be part of a major holiday, so you might as well pack as much quality time into that holiday as possible. Relaxing with friends at a café or bar, overlooking some spectacular scenery, can be most enjoyable, as depicted in Fig. 8.1. Painful though it may be to admit, your eclipse might be clouded out. If the eclipse only forms part of your holiday then it will not be such a great loss. Veteran eclipse chasers tend to travel with good friends and with the same travel companies, time and time again. In the company of such friends you can recount the good and bad and downright bizarre experiences of previous trips. I find my memories of specific totalities tend to be rather vague and merge with other eclipse memories. However, unique and hilarious events stick in my mind and become great talking points at the bar on the next trip. Travel to the so-called "third world" countries is always an experience for the wealthier western tourist. Travel broadens the mind, especially when it is to somewhere unique, different and not as comfortable as home. It makes you appreciate a civilised country all the more. When an eclipse occurs across a third world country, literally anything can happen. An infrastructure that is fairly fragile anyway suddenly faces an influx of 10,000 people; they all want to get to a thin eclipse track, stretching across the country. Airport services quickly become overstretched (this even happened in Hawaii in 1991) and hotel rooms can often end up being double or triple booked if you are not with a reputable company. These events can seem like a nightmare at the time, but they produce endless stories to recount on boring airline flights, coach journeys and evenings at the bar on the next holiday. Similarly, the cranks who attend all eclipses can generate much amusing banter for future holiday trips. However, sometimes the horror stories are a bit close to terror for comfort. My friend Nigel Evans went to Turkey for the 11 August 1999 eclipse. In his own words:

On the Tuesday morning after the eclipse, 17 August, we were in Canakkale and I dreamt that the room shook. When I got down for breakfast I discovered that

Fig. 8.1. Eclipse chasers Nigel Evans and, in the background, Richard Monk relax in a Santorini bar after the 29 March 2006 Explorers Eclipse Cruise to Libya. (Their Cruise ship, the MV Perla, is the nearest ship in the background.) Image by the author

I hadn't dreamt it. We were about 160 miles from the Gulf of Izmit, the epicenter of a magnitude 7.8 earthquake at 3 a.m. local time. Once we realised how serious the earthquake had been we all phoned home to let everybody know that we were OK. We then journeyed via the European side of the Sea of Marmara to Istanbul, which is about 50 miles from Izmit. We saw no signs of damage, just people camping out on the roadside. In central Istanbul, everything was reasonably normal, except that many tourist venues were shut. Early the following morning on the way to the airport, we saw thousands camping out. Over ten thousand people had died.

To add to any eclipse experience it is highly advisable to research the area you are going to before any journey. Sometimes you are within a few hundred kilometres of a real "wonder of the world" and it is tragic if you do not visit such a place on your (quite possibly) only visit to a country. Eclipses have shown me some spectacular scenes which I simply would never have got round to seeing if not for totality tracking across a foreign land. On my various astronomy travels I have witnessed (amongst other things) the spiritual environment of Machu Picchu in Peru, the barren wilderness of the Atacama Desert in Chile, the Taj Mahal in India, the tea plantations of Darjeeling, the big island of Hawaii, Victoria Falls in Zimbabwe, the Roman Ruins at Leptis Magna in Libya, and the spectacular island of Santorini in the Mediterranean. In addition to these sites I have learned to Scuba dive in the Caribbean, continued this experience in the Red Sea and spent 9 days on a cruise ship. These are all destinations and experiences that I simply would never have found good enough excuses to visit if not for eclipses and Leonid meteor storms. At the time, some of these trips involved quite a bit of hassle, but, looking back, they were all memorable experiences, and surely the best purpose of this painfully short human life is to "experience" as much as possible. Indeed, nothing can truly be experienced unless you are a sentient human being with the ability to store memories. None of us would be anything if not for our memories and experiences; we would all be like babies with nothing to relate to. Experiencing events in the way we do may be unique to human beings (Okay, maybe Dolphins too!) so what great fortune that us humans live on a planet where the Moon and Sun are the same size in the sky and total solar eclipses occur regularly enough to see many in a typical lifetime. For me, on a personal level, eclipse chasing *is* the nearest this devout atheist can come to having religion. Sharing these experiences with like-minded and tolerant friends is a major part of the whole eclipse-chasing experience and I can honestly say that the funniest experiences of my life have come on eclipse trips. When I see large groups of football supporters travelling the world, just to savour the possibility of that one big conquering win; or a hat-trick of goals from their favourite striker I cannot help thinking that eclipse expeditions are rather similar. Totality is like the killer goal and cloud is like the worst possible defeat. But everyone is in it together and will relive their memories of the big day for the rest of their lives.

I would now like to take up a few pages to share with you some of my most vivid, bizarre and surreal eclipse experiences.

Some Unique Individuals

On my trip to the 3 November 1994 Chile/Peru total eclipse we had a German traveller on our itinerary, who had a staggering alcohol requirement. He was travelling (unknown to his wife) with his mistress who was constantly apologising to others for his "problem". There were quite a few internal South American flights on that holiday (all characterised by an onboard Bingo game!). On one such flight, as the plane took off, the German in question got up from his seat and, with the plane climbing steeply, trudged uphill towards the cockpit (imagine that happening in a post 911 world in the USA!). The amazed stewardesses intercepted him just before he reached the cockpit. When asked what he was doing he explained he was "going to the bar for a drink" and, promptly entered a lavatory cubicle. Well, I suppose people drink all kinds of liquid these days. When the plane landed our colleague half-stumbled and half-fell down the airline steps, but he dusted himself down and marched on regardless after he picked himself up. It was like watching a Buster Keaton silent movie. But the best was yet to come. A few hours later our group arrived at a nature reserve for condors. Suddenly a group of local Peruvian children ran towards us bearing postcards. We all did a double-take: they were postcards of us, arriving on the plane, earlier that day. Then we spotted they had done a special presentation set of our drunk German colleague staggering down the steps. No one was interested in their own mugshot postcards after that. Large sums of money changed hands for examples of that one set of postcards. We were laughing so much we were in physical pain and none of us had been drinking. (Unlike our German colleague, who was still drunk but was not enjoying himself.)

Travellers on the 1991 eclipse expedition to Baja were fascinated to see that a very famous German eclipse chaser (no, a different chap) had engineered a very strange-looking wooden mounting for his celestron "Comet-catcher" telescope, despite not having brought a mount with him. In fact, the mount had been made by utilising the large bedroom drawer from his hotel room which was just the right size when physically broken into its constituent parts. The description of this apparatus at a subsequent astronomy meeting in the UK had people, again, writhing in pain from the audacity of the carpentry.

On my trip to the clouded out Hawaii eclipse, in July 1991, I was walking along a road in Anaheim, Los Angeles, with a fellow amateur astronomer. It was just after the time of the first Gulf War, called Operation Desert Storm, and the US Military was being moved out of Kuwait, Saudi Arabia and Iraq. In retrospect, straying down a poorly lit road in Anaheim was probably not a good plan. Suddenly, we spotted a tall young man, in military uniform, moving towards us. He had a strange demeanour which immediately gave me the creeps; a bit like some character out of a Stephen King novel, or the liquid metal terminator in *Terminator 2*, played by Robert Patrick. He told us, in a precise and measured way, that he had just come back from his patriotic duty in the Gulf and been dropped off by the US Military, on that road, but without any money or accommodation. His manner was robotic and calm and his steely gaze was fixed on us. I didn't much like the bulge in his military jacket either. He repeated the fact that he needed money, and that he was a patriot. His expression never changed. After many minutes we seemed to convince him that we, ourselves, had only just landed and were in the same situation, i.e. we had no dollars, just a few useless British coins. I gave a huge sigh of relief when he finally glided away. On the news the next day a drive-by shooting in that exact same area was reported. I couldn't help thinking that the mysterious military man had acquired a car. Not one of my favourite eclipse trip memories. I have felt much safer in the most deprived third world areas I have visited than in that part of Anaheim.

To reduce the expenses of a holiday, younger eclipse chasers often take a gamble and decide to share rooms with a complete stranger. This option exists on many eclipse holiday booking forms for same sex (does that have any meaning these days?) and compatible smoking habit travellers. However, do it at your peril. Even good friends (or newly weds) can irritate the hell out of each other on holiday, so the potential nightmare of sharing with a stranger is almost infinite. One traveller on an astronomy holiday discovered that his fellow traveller liked to wear a balaclava indoors all the time (and a crash helmet sometimes). Also, when the sane traveller was lying, having a soak in the bath, his balaclava-wearing sharer would walk in, stark naked (except for the balaclava) and have a dump in the bog in full view of him. You do not want that in your eclipse-travelling life. Sharers who not only snore, but, and I quote a friend "snore from both ends, at 200 db for eight hours" can make you want to shoot them, or inflict their torso with a thousand stab wounds, as they sleep. Okay, you would go to prison for life but it would be worth it though. Think *very* carefully before sharing with a stranger.

Some eclipse experiences have been total and utter nightmares for the individuals concerned. I will not mention the name of the elderly lady who, tired out by the travel and jet lag, decided to get a good night's sleep before the pre-dawn rise at the eclipse site, by taking some powerful sleeping tablets. Yep, you guessed it, she slept right through the eclipse, and the tour guides hammering on her dead-locked room door to wake her up and get her on the coach. Perhaps even more terrifying was the experience of a traveller on the 10 July 1972 Atlantic eclipse cruise (the track passed over Nova Scotia and ended in the north Atlantic). The eclipse chaser, who I will refer to as "Mr S", was happily set up on deck, with his equipment, half an hour before totality when he decided to just nip down to his cabin and fetch an extra piece of kit. While down there, the ship's captain turned off all the ships lights (as previously arranged by the tour, for the benefit of the eclipse chasers) and Mr S found himself in complete darkness (there was no mandatory automatic emergency lighting requirement in the 1970s). He fumbled around blindly trying to retrace his path as the clock ticked inexorably towards totality. Eventually he made it back to the deck just as the third contact diamond ring flared signalling the end of the eclipse. Presumably Mr S needed professional counselling for the rest of his life.

Some Unique Countries

India

I have travelled to India twice, once for the 1995 total solar eclipse, and once for potential viewing of the 1998 Leonid meteor shower. Both experiences are indelibly welded on my mind. Once you leave the air conditioned environment of Delhi airport you are hit with a solid slab of humid smelly air, with a distinct and unforgettable smell. The inimitable John Mason, tour guide on all *Explorers* astronomy holidays summed it up perfectly: "A delightful aroma of raw sewage, wood smoke and diesel fumes: Lovely!" The total eclipse shadow track of July 2009 and the annular track of January 2010 cross India so you may be heading there soon. India is a country you can love and hate simultaneously. You hate the stomach upsets, the smell, the sheer volume of people and being flocked by locals keen to sell you junk for western currency. You love the way the people simply survive by bodging everything to work and fitting three or four family members on a scooter, a dozen people in every car, or thirty people on the back of every truck. Do not be surprised to see people cycling the wrong way down the motorway fast lane or to encounter a 4 cm high tarmac step on that same un-illuminated stretch of motorway at night. Do not expect drivers to own working car headlamps or a rear view mirror either. Painting a colourful message on the back of a lorry saying "Sound Horn Please" is cheaper and more traditional than using mirrors, and who needs mirrors when the entire family is looking out of the back of the cab anyway. I have seen bus drivers doing a three point turn on a roundabout in India and a small child weaving a rickshaw laden with gas cylinders labelled "Pure Hydrogen – Highly Flammable" through a roundabout in the Delhi rush hour.

Many hotels in India are called "Ashok". I am convinced this is partly to do with the electrical wiring. In one hotel I stayed at in 1995, I was thoughtfully provided with matchsticks to wedge my mains plug tightly into my mains room socket. I had brought a complete set of international electrical adaptors with me, but did not realise that the pin spacing and the socket hole diameter can vary widely from hotel to hotel. Also, do *not* make the naïve mistake of imagining that the lights in your room are operated by switches in your room. Finding out which vital switch connects to which light is usually a challenge in most hotel rooms; indeed, even the intellect of Johann Carl Friedrich Gauss would be stretched to decipher the logic underlying the lighting or shower control symbols in the average third world hotel room. These symbols may mean something to the designers, but they mean diddly-squat to the tourist. But at least, outside India, the switches are usually in the same room as the lights. This was not so on the 1995 eclipse trip. One colleague found that half his room lights were operated from the room next door, necessitating the occasional bang on the wall and a shout to his amused neighbours, who were happy to oblige, provided they were not otherwise engaged in their own banging activities. Another colleague on the 1995 eclipse trip innocently switched on the hot water immersion switch to the boiler in his room and there was an explosion, which, and I quote: "Made Krakatoa look like a chip-pan fire." Fortunately no one was injured.

Water is valuable and expensive in India and many hotels do *not* provide bath plugs for this reason. The wily eclipse traveller will bring a set of various diameter bath plugs on holiday to cope with this. I find it somewhat disconcerting when, despite a big sash labelled "Sanitised for your convenience" across the toilet bowl, flushing the toilet produces brown smelly water! Maybe I am just a cynic? Numerous sources of income, based on bureaucracy to milk the tourist, exist in India and many other countries. Someone once said that the British invented bureaucracy and India turned it into a science. I think I would re-word that as "a money-spinning" science. When a thousand eclipse chasers descended onto the ancient fort of Fatephur Sikri, not far from Agra (and the Taj Mahal), in 1995, the local authorities suddenly invented a tripod tax. An old Fakir (yes, you can pronounce it any way you like!) was posted at the fort entrance demanding a tax from the hordes of British, American and Japanese tourists. Not surprisingly, as he had no firearms, was small, old and bent, and we had loads of weapons to threaten him with, we ignored him. The situation was somewhat different inside the fort. Our travel company had acquired written permission to use an entire section of land within the fort for our cameras, tripods and equipment. Unfortunately, the Japanese contingent had arrived there just before us and paid off the Indian military to guard their land, armed with rifles. It made no difference that we had paperwork to support our claim. For once official paperwork made no difference to the Indian military, because money talks with more authority, as do the rifles of the Indian Army. As one happy smiling and friendly Japanese eclipse chaser told us "Dey have reen toad to krill you stone dead if you take our space! Grame On!" So, we were resigned to the cramped upper stories of the fort, just above the Japanese observers. Of course, gravity is very good at carrying rubbish and a variety of yellow liquids downwards, not upwards, so maybe we did end up in the best spot. We also learned that the serene and contemplative Japanese eclipse way is to announce every eclipse stage with a tannoy system, blaring out the minutes and seconds to totality.

After the 1995 eclipse a post-eclipse buffet was held on the next night, with standard Indian cuisine. A friend of mine told me the salad was very good and the ice cream was fantastic. I stuck to my guns: I was not going to eat anything in India unless it was cooked to a crisp, however tasty it looked. A day later, at Jaipur, one end of him was welded to a toilet, while his head was positioned over a sink. He was in a similar state on the 6-*hour* (with one stop) bus ride to Delhi. So were other colleagues, decked out with rectal bungs and home-made nappies. I almost said "I told you so" but he was a good friend in a pitiful state. Never eat uncooked food in India unless you are in the finest hotel and a few hours from returning back to the UK and can be ill in your company's time, not *your* holiday time.

Quite a few American and British sitcoms, films and dramas can be viewed in most Indian hotels and frequently the soundtrack is replaced by a local Hindi one and may be subtitled back into "almost-right" English. Often the hotel may produce an English version of the TV schedules for western visitors and this can have you rolling around in hysterics when you read it. I will give some examples of the programs available "for my delight" in my hotel Room in October 1995. Two films stick in my mind: "Bitch Cassardy and the Samdance Kid" and "Ravange of the Kaller Tomatoe." Then, to round the evening off, a British sitcom: "Groege and Mildred".

The airports in India are, perhaps, the areas where your patience will be tested to the absolute maximum. Even prior to "911" anything containing batteries were regarded with great suspicion, and yet, if a camera cannot be proven to work, by demonstrating the battery-powered flash going off, it may be confiscated. If you have any answers to that dilemma, send them to me, on a postcard, please. Any loose batteries can be confiscated, depending on the whim of the "Jobsworth" official on the day, prior to boarding a plane. On arrival at the other end, the batteries can be reclaimed if you have the necessary paperwork. Invariably the batteries you are given back will be similar, but older than the ones you handed in and, mysteriously, the most expensive batteries will be missing. Thus, carrying spare batteries everywhere, in your luggage, is a good idea: They are happy when you hand over a token battery but they do not search your entire luggage; so you are equipped to suffer a 25% battery loss. Remember, these "Jobsworth" officials have rare, salaried jobs with a pension. Their prime motivation is not common sense, but keeping their job by doing everything by the book, however bizarre. They also like the sense of power the uniform gives them. You simply have to smile at them, flatter them and humour them, even if you would really like to bang their head into a wall.

A similar battery/camera obsession was experienced by me, not in India, but at the Peru/Chile border crossing at the 1994 eclipse. I was in a minibus containing half a dozen eclipse chasers, including the Queen guitarist Brian May. We must have had three film cameras each and one video camera each. The authorities at the border announced they wanted to tick off every single camera that crossed into Chile in case it was an illegal import and needed tax paying. Fortunately, Brian convinced them, in fluent Spanish, that as we were overloaded with cameras this would take all day. So they just limited the declaration to video cameras and confiscated Brian's Peruvian oranges instead. Once our video cameras were declared they issued us with tiny slips of easily losable paper which we had to surrender on re-entry to the country. Of course, on leaving Chile a few days later the authorities wondered why only video cameras were on the chits of paper and not the film cameras. It was all a very random process, done at the whim of the "job-for-life" rubber-stamping gun toting officials, but as we were in a large group with some good, experienced, travel reps on the way back we had sheer quantity of numbers on our side. While waiting at the border crossing, one character, in a large white American car, was having his car searched with the boot open. While the guards were distracted he made a run for the Chile border, by starting the car and flooring the accelerator pedal. He chickened out of crashing through the Chile barrier though and was promptly arrested.

Bear in mind that if you purchase any foreign currency at an airport like Delhi you will get a cubic light-year of the local currency in exchange for your western currency. The massive wad of Indian rupees will inevitably be stapled through by a humongous staple and in the airport cafeteria, a giant staple extractor can be hired for a fee (Fig. 8.2); at the same time your pot of tea will be delivered complete with quadruplicate forms to sign for accepting the merchandise. If entering the streets of any major Indian city wearing western clothes, all eclipse chasers will be herded into the nearest carpet emporium where smartly dressed Indian men wearing fake Rolex washes will tempt you into buying a Kashmir carpet.

Without doubt the most bureaucratic airport I have visited on any astronomy holiday was Bagdogra which is the gateway to the road to Darjeeling where I headed for the 1998 Leonid meteor shower. A typhoon headed up the coast of India with us, almost unheard of in November in Darjeeling. At every internal stop on the route to Bagdogra the same plane steward asked every remaining passenger to identify every item of luggage in every overhead locker. Just what you want when you have been travelling for 36 h! On arrival at Bagdogra, sweaty and dog-tired, you are then ushered into a holding area where everyone's passport is seized (a disturbing development at the best of times), a mass of forms are produced for

Fig. 8.2. Eclipse chaser Dr. John Mason attempts to extract the industrial strength staple from his wad of Indian rupees. Despite pleading with the Delhi airport money changer the staple was inserted as per the unbreakable regulations. Minutes later the airport café loaned a cutting implement from the cutlery drawer for the staple's removal, so the cup of tea could be paid for in the local currency. A typical example of one of the bizarre bureaucratic hazards of eclipse chasing in third world countries! Image by the author

every traveller to sign, and the sound of rubber stamping can be heard for hour after hour with your tour rep being frequently summoned to answer questions. The eclipse expert, meteor expert, and tour guide, John Mason, told me that on a previous visit bureaucracy had been anticipated, and so a meticulous spreadsheet printout had been prepared with everyone's address, passport number and other details already filled in. Helpful eh? Nope! They were told that the meticulous form was in landscape format which is illegal; only portrait format is acceptable for such details.

Some 5 h later we were in a minibus with all our vital equipment bunjy corded to the roof and some 40 h without sleep. We were well on the way to Darjeeling and were crossing a rickety wooden bridge, with a few gaps in the structure, overlooking a ravine. Our way was blocked by a wide lorry coming towards us as another lorry overtook us. All three vehicles ground to a halt and there was no obvious way out; the situation was made worse by the fact that it was night-time and the road was in darkness. This may seem bad enough, but this was only the start; read on!

The oncoming carriageway was not only a road; it also incorporated the rails for the local train. Somewhat predictably, with the carriageway completely blocked, the whistle of the train was distinctly heard and the locomotive soon hurtled into view. I will never forget John Mason's shriek in the darkness. "Well, that's all we bloody need!"

I have never seen two lorries get out of an impossible situation more quickly, one choosing to reverse back past us (and presumably plunging into the ravine) while the other squeezed past us with inches to spare with the train on its bumper.

Had I ever been closer to death? To be honest, after about 40 h with no sleep, I almost relished that prospect.

Zimbabwe

My overwhelming memory of my 2001 eclipse trip to Zimbabwe was of being separated from my friends and ending up in hotels where I knew nobody. How did this happen? Well, I was on the same itinerary as most of my friends *but*, due to the volume of people travelling on that itinerary I was separated from them. We were split amongst numerous hotels. This is a lesson for all single eclipse chasers. The only way to stick with your friends is if you book the holiday as a family group on the same form, or if you are on one big cruise ship. Being in separate hotels five miles apart is not much fun. Also, fully verify that your mobile phone and your friend's mobile phone will work in the country you are headed for. We had been told on the trip that the eclipse site was near a school and so if we brought pens, pencils and notebooks as gifts they would be very much appreciated. One of my most vivid recollections on any eclipse trip was opening my bag and seeing the children's faces as I pulled out some of these items. It was as if you were handing someone a cheque for thousands of dollars. They were deliriously happy, and most of the travellers around me had done the same. My funniest recollection from that holiday was the effect the deluge of western credit card owners had on the restaurant employee at the hotel in Johannesburg, just prior to our short flight into Zimbabwe itself. The hotel was obviously not normally geared up for so many people paying with plastic, but that was only half the problem. The ancient keyboard terminal had a dodgy number "8" character on it (which was tough if your expenditure had a number "8" in it, and it took 30 min for each transaction to be validated by the ultra-slow and dodgy modem link. Funny the things you recall on eclipse trips. Eclipse day in Zimbabwe (21 June 2001) had its own dramatic memories too of course, not least the vocal comments of a first-time totality witness, standing a few yards away from me who was totally staggered by the site of the eclipse. His precise words as second contact occurred, taken off my video tape, were

> Ohhhhhh My God, I can't stand it. Look at that! Bllllooooodddy Hell. Oh My Lord, Oh My God, that is staggering. I'm blessed to be alive. My Lord, that is fantastic. My heart is going like the clappers. That is like the slowest sex you can imagine. . . . . . Hubba, Hubba, Hubba . . . .

Well, eclipses affect different people in different ways!

Cruising Along

Without doubt a holiday on a cruise ship is the most luxurious way of seeing any total solar eclipse. There are many advantages. The biggest ships can accommodate over 2,000 passengers so there is no risk of being separated from your friends. The onboard facilities are often so comprehensive that it is like sailing along in a shopping center. A ship can head for the clearest patch of sky along the shadow track in the days before the eclipse, with no extra customs and immigration hassles in international waters. In addition, every night, even if you have had a day's excursion into a foreign country, you return to the same room for the duration of the trip. On so many whistle-stop trips I have been on, travellers have a different hotel room every one or two days, which simply leaves them head spinning. It takes at least three days to work out how the facilities in each bedroom work. Another advantage of cruising in today's modern world is that digital images and video films of the eclipse can be combined into an impressive *Powerpoint* presentation on the way back, after the eclipse, to relive the experience in the ship's lecture theatre. A further fascination is that you can all crowd around the deck to hopefully see the so-called "Green Flash" as the Sun sets over the ocean (see Chap. 5). Finally, for me at least, there is something quite extraordinary about sailing through a big sea in a huge ship; at night it is almost as if you can imagine being on a spaceship and, of course, in the middle of the ocean, there is no light pollution apart from the ship's own lights. Of course, if your eclipse track does not cross over the sea then you have a big problem. However, the majority of eclipse tracks do pass over the ocean and some of them, as we have seen, pass exclusively over oceans. Then there is the matter of seasickness. In a calm sea (like the Mediterranean) and on a reasonable size ship (more than 10,000 ton), most people will just feel a gentle swaying sensation. In choppy seas on such a ship seasickness will result, but many ocean-going cruise liners have displacements many times this size, up to *and over* 100,000 ton in some cases. It takes quite a choppy sea to make a trip on such a ship unpleasant and, of course, the captain will endeavour to avoid such conditions.

Let's get one thing straight, eclipse cruises attract cranks like a compost heap attracts flies. The cruise ship I travelled on (the *Louis Lines MV Perla*) for the 2006 March Eclipse was full to the brim with them. Religious weirdoes, men be-decked with jewellery, men dressed entirely in red, men with pony tails, men who stammer like a machine gun and then masticate like a pair of castanets, men who may well be women, and I won't mention the hair coloration, just take my word for it. The ship was sailing with a full crew, but that is more than could be said for some of the passengers. People who I had known from many previous eclipses seemed to be knitting with one needle on that trip. Perfectly sane *male* individuals arrived on the ship with hair painted green and adorned with more gold jewellery than the *A Team's* "Mr T" on that ship.

One phenomenon that everyone was interested in onboard that ship was spotting the "Green Flash" as the Sun set in the ocean. However, after spending the day before totality (Tuesday) at the old Greek and Roman ruins at Cyrene in Libya, and experiencing the horrendous state of their lavatories a new astronomical term entered the language, namely the "Brown Flush" and no, the brown colour was not due to rust or soil! The lavatories were blocked *before* we arrived and most of our 800 people were desperate to use them. Flushing them brought up more waste then went down – nice!! John Mason, who was the tour guide on bus number two, was going around the site collecting his people. As he passed near the lavatory queue shouting "Any more for number two" one wit shouted back "You wouldn't even risk a number one in there John". Expect the most horrendous lavatories imaginable in third world countries and use a bush instead. The outside undergrowth is far more hygienic.

Buses and Bus Drivers

I have already mentioned some experiences in India while travelling on the roads there. However, most of the poorer countries have some pretty unusual buses, drivers and highway-code systems. For land-based eclipses you will, inevitably, find yourself travelling in a huge bus convoy, quite often with a police or military escort to protect you from any opportunist thieves. Thieves are only too aware of the vulnerability of travellers luggage when it is left on the kerb prior to loading or when being removed from a bus. Indeed, they took full advantage of this on the 1994 eclipse

Fig. 8.3. The name of the bus says it all. A coach used on the 2006 Explorers eclipse trip to the Libya desert. Image by the author

where various travellers in Peru were distracted while thieves stole items of baggage. On the Libya eclipse our driver was named Mustafa. Bearing in mind his stuntman overtaking manoeuvres, I assumed his surname was Naxident. Overtaking on almost blind bends, while driving a bus full of passengers, seems to be a perfectly acceptable practice in countries like India, Chile or Libya, and it will alarm the more elderly first-time travellers on these trips. There seems to be intense competition amongst bus drivers too; even when driving in a convoy the question "who has got the fastest buses" seems to arrive. Never is the competition fiercer than when bus companies and bus drivers from neighbouring countries are involved. A sight I will never forget on the bus trip into the Libyan Desert is the two drivers changing over. What's odd about that you might ask? Well, they did it while driving! Set the cruise control up and just stroll out of your seat, walk into the spare front seat, while the relief driver casually takes his place. I heard one traveller saying that their drivers had done this switch *while overtaking*; although as he'd had a lot to drink at the bar I think it was just a story! Or was it? Still, the buses we were travelling on were owned by the "El Joker" travel company (Fig. 8.3), so I suppose we should not have been surprised.

Dehydrate or Risk Your Bladder Bursting?

I have already mentioned lavatories. The eclipse chaser in a hot (aren't they all?) third world country often faces a dilemma on long bus journeys. Drink plenty of bottled water as advised by the tour reps, and then spend the whole coach journey waiting for the next "comfort" (Hah!) break, or don't drink much water and risk serious dehydration. For the elderly it is a real dilemma. After facing this knotty problem on a number of foreign trips I now drink lots of water the night before and then don't drink any on coach day until I start feeling really thirsty. I also avoid exertion and Sun as much as possible. Foreign lavatories, as we have just seen can be a nightmare from hell. A woman on the Libya eclipse trip followed the advice to drink loads of water and then sprinted hundreds of yards into the desert, when the buses stopped, to relieve her bladder before she detonated. After a cubic light-year of her bodily fluid had been emptied into the sand she promptly got up, slipped on the sand and rolled over in what she had done. Fortunately, I was not sitting next to her on the hot coach journey back!

Finally in this chapter, bear in mind that many countries issue commemorative eclipse stamps when a total solar eclipse passes through their territory. They make a nice memento of the event alongside the photos in the eclipse album (see Fig. 8.4).

Fig. 8.4. Eclipse stamps (in this case for the Libya 2006 totality) are invariably produced by countries experiencing totality and become collector's items for many eclipse chasers. Image: Andrew and Val White

Chapter 9

Checklists and Travel Plans

I have learnt a huge amount about how to enjoy astronomical excursions abroad, ever since my first trip to Tenerife, to see comet Halley, in 1986. What item would I place at the top of my packing list? The answer may surprise you. It is a pair of wax earplugs! I envy people who can sleep on a whim when they are tired. I cannot sleep, even when dog-tired, unless I am comfortable and there is zero noise in the room, hence the earplugs. As for people who can sleep on a whim, and snore like a 150 db road drill, thereby happily (and smugly) preventing others (even in the next room!) from sleeping, well, the electric chair is much too good for them! I have tried various types of earplug, but wax ones mould themselves to the ear-hole shape and you cannot hear a thing. Okay, you can't hear the hotel fire alarm either, but I would rather risk burning to death than have an entire holiday with no sleep. Second on my packing list is simply my list itself, because as well as just a list of things to take it contains crucial reminders of what equipment to take on eclipse day and what procedures to adopt to avoid hassle. Most people who are on a holiday in a far-flung location will feel like their head is spinning, with their daily routine disturbed, jet-lag and simply remembering all the little bits and pieces they have to carry with them. A checklist means you can put less strain on your creaking grey matter and enjoy the holiday more.

The Big Day

At some point, before the holiday, you *must* take every item of eclipse equipment outside, into your backyard, or back garden, and test it as if you were actually doing it for real on eclipse day. That is assuming you are not just going to sit back and watch it. The equipment for the partial phases should be tested on a sunny day outside and the equipment for the total phase should be tested on the Full Moon at night. At some point you will find yourself in a desert somewhere (see Fig. 9.1) far from home and vital spare parts, wishing you had made better preparations. Now, we are all a bit lazy when it comes down to carrying out rehearsals like this. It is just so much easier to throw everything into your case and assume it will go well on the day. Take it from me, you *will* regret it *big time* if you do not have a dry run. Also, during your rehearsal, carry a notepad with you to note down the little hassles you encounter; things that need to be fixed before eclipse day. All this might seem a bit paranoid before you have been to a TSE, but once you have

Fig. 9.1. The author, on site, at Jalu, in the Libyan Sahara, on 29 March 2006. Image: Nigel and Alex Evans

Fig. 9.2. Get familiar with your equipment from the business end. A Canon 20D attached to an awesome Canon EF 500 mm f/4L lens, a superb combination for the eclipse photographer. Image: Jamie Cooper

experienced the sheer panic of those brief minutes of totality you will know where I am coming from. Eclipse equipment needs to be slick to use, intuitive and 100% reliable, and there must be no sudden nightmares. Quite possibly you will be setting up your equipment on sand, with your skin covered in slimy sunblock (dripping into your eyes; been there, done that; a nightmare!!), and with a brisk breeze trying to whip your solar filters off. You will quite possibly be very hot and sweaty too and fighting off a plague of insects. You need to know and feel exactly how ergonomic your proposed set-up will be, staring at your camera back (Fig. 9.2) and its control panel at, quite possibly, a neck-breaking angle, and in semi-darkness. Everything you use in those panic-stricken minutes (or seconds) of totality must be tried, trusted, and ergonomic. Cameras and video cameras on the same tripod head should be precisely alignable with ease, a prominent clock should be visible (or quietly audible) and your cameras should be movable in smooth slow motions (not panic-stricken jerks). Plus, all this has to work in the darkness and solar filters have to be "whippable offable" with zero hassle at second contact. Don't say I didn't warn you!

If It Can Go Wrong . . .

Quite often you will encounter unexpected problems on the big day that only the most paranoid planning would have allowed for, so you need a bit of adaptability on the day. Here are a few things that I really had not expected on my own eclipse trips.

1. In Hawaii in 1991 I really did not expect it to be so cloudy. Prior to the trip everyone had said that good weather was a dead certainty. Afterward everyone said Hawaii was always cloudy at ground level in July! Who was right? I have no idea. Neither did any of us expect to be with a tour guide who would not let our coach ascend the slopes of Mauna Kea. If he had, like the other tour guides, who had a friendly word with the police manning the road blocks, we would have seen it. Sadly, our tour guide was incapable of having a friendly word with anyone. It was 3 years before I recovered from that depressing experience. After that nightmare I always look at whether the next eclipse crosses a proper desert and, if so, I head for the desert! Nowadays I would also consult the wonderful Solar Eclipse Mailing List (SEML) at http://tech.groups.yahoo.com/group/SEML/, moderated by Mike Gill and also at Jay Anderson's weather prospects site (http://home.cc.umanitoba.ca/~jander/). Deserts are almost always the best places to be for an eclipse. By definition they have little rain (or snow) and very little cloud either. When thinking eclipse, *think desert*!

2. In Chile in 1994 altitude sickness gave fully half the people on the trip a splitting headache on eclipse day. The military coaches were powerful but had no air-conditioning and almost no suspension. The ride back was truly an endurance expedition and my packet full of chewable aspirin was an absolute life saver.

3. As I mentioned earlier we did not expect to be forced onto the upper ramparts of the fort at Fatehpur Sikri, in India on eclipse day, in 1995, but then we did not expect the Japanese to hire armed protection! Because of the limited space my tripod had to be extended to its full height, resulting in some very shaky pictures. With seconds to go to totality a *Discovery Channel* cameraman muscled in to our patch trying to get a face-to-face interview with colleagues. He almost wrecked our experience and we let him know. TV people near you can seem quite glamorous on eclipse day. Ignore them! They do not care about you, the people around you and you will not end up famous. They are after one thing: video footage. The chances are that their film will make eclipse chasers look like weirdoes and they will only use a tenth of the film they shoot. Give yourself a buzz and tell them to shove their cameras where the Sun is always eclipsed; or, ask them how much they are paying you and then tell them where to go. Above all, let them know that if they wreck your experience of totality they will

seriously regret it. Loads of amateur astronomers seem to crave a bit of TV fame. Then when they see how silly they are portrayed on TV, they are much wiser.

4. A colleague's wife was attacked by a sand fly on the beach at Bonaire just after the Caribbean eclipse. Her leg was bitten and literally turned into a bloated and swollen cylinder of throbbing pus! Needless to say, we all took pictures of the phenomenon! Be alert to these little horrors and wear insect repellent sunblock.

5. The 1999 Cornwall Eclipse was cloudy. But then what do you expect in England! However it was surprisingly eerie under all that cloud and still a fascinating experience for those of us who had seen totality before. The eclipse went wrong, but, unusually, it was still an unforgettable experience and no-one had travelled that far to get there.

6. The terrain at the Zimbabwe eclipse site comprised the finest sand imaginable. It got everywhere, including inside cameras. The only alternative was some very slippery rock, but both surfaces were very tripod-unfriendly. If you really want a dress rehearsal for eclipse day, do it on a beach. To prevent others coming near your equipment you may like to consider the "Evans" technique, practised by Nigel Evans at many eclipse sites. Simply get as near to the water as possible, facing the Sun, so no one can get in front of you (see Fig. 9.3). Of course, if you get your tidal calculations wrong you may end up underwater before totality!

7. We had thought that allowing 7 or 8 h for a 500 km bus trip through a decent road in the Libya Sahara in 2006 would be more than enough. Yeah, right!

Fig. 9.3. The hours before first contact can be quite frantic. Nigel Evans rushes to set up all his cameras and hopes that the tide does not come in! Picture taken by the author at Knip Bay, Curacao, on 26 February 1998.

It took 9 h. Of course, we had assumed the 23 bus convoy would have filled up with diesel the night before and that a major gas station in Benghazi would have more than two working pumps to supply 23 coaches. Then we had not realised that a convoy would be enforced rigidly, so that all the buses would stop at the same lavatories in one go and 800 people would queue up to use two toilets!! We just made it to the eclipse site in time to set up the equipment!

Baggage Allowance

Most eclipse travellers push their luck where baggage allowance is concerned. The standard allowance on most UK departure eclipse flights I have been on is 20 kg for checked in luggage and 5 kg for carry-on baggage (1 kg = 2.2 lbs). A limit on the size of the carry-on baggage is typical too; even more so at times of high terrorist fears. In the hot summer of 2006 UK airline security measures temporarily reduced the onboard baggage allowance to virtually zero. So, in these troubled times, anything can happen. However, in normal circumstances UK airline officials seem to turn a blind eye to check-in baggage up to 25 kg and whether they weigh carry-on baggage seems to be totally random. It should be explained that, in practice, they are not worried about the total weight of the carry-on baggage, but whether the largest chunk is over 5 kg. At some point some "jobsworth" official in Europe has decreed that anyone being hit by a piece of luggage, over 5 kg in weight, falling from an overhead locker, will be seriously injured. But at 4.9 kg they will not feel a thing! So if you have all your expensive camera gear in a hefty carry-on bag you may have to remove various cameras into separate bags at the check-in, until the weight of the main bag is less than 5 kg. Most eclipse flights are full of travellers hauling carry-on baggage (stuffed with cameras) that is over the weight limit. After a while the check-in staff can get really tired of asking everyone to remove a camera or two to get the weight down, and they just give up. Otherwise the people will still be checking in at take off time! So although it is nice to be at the front of the queue and, maybe, negotiate a better seat next to your friends, being twentieth in the queue on eclipse flights is better than being first sometimes. One thing I have noticed in the past is that if you check in last, your luggage often comes off the conveyor first at the other end. Typical European baggage allowances make life quite tricky for the eclipse traveller who may wish to take an equatorial mounting and some small telescopes or large lenses on the trip. If you are travelling with a wife, partner or friend the camera equipment can be spread between two cases but for the single traveller a maximum of 25 kg is quite restrictive. However, most European airlines do exclude laptop PCs from the baggage allowance, so at least these can be carried without affecting the weight restriction. Looking smart, well shaven and polite and friendly in appearance can still get you a sympathetic ear, even in these worrying post-911 times.

On some flights a reasonable amount of excess baggage can be carried if an extra fee is paid. In the USA, especially on internal flights, these allow considerably more weight to be carried onboard. Typically, two items, weighing less than 45 kg, with the heaviest item weighing less than 32 kg is allowed. However, even these excess weight allowances seem to be dropping in recent years with individual items over about 23 kg being banned in some cases. This virtually eliminates

Fig. 9.4. Some eclipse chasers think big! A 35 cm aperture Schmidt-Cassegrain was taken to the 2006 eclipse in Turkey by one observer. Image: Jamie Cooper

large telescopes being taken abroad. Having said this, at every eclipse site there is always someone who has managed to transport a 35 cm aperture telescope along (see Fig. 9.4).

However, it is essential to get the facts, for your flight, beforehand, and establish exactly what rules apply in your case.

Checklists

I must admit to being an obsessive–compulsive list maker. I just hate arriving in a foreign country and suddenly finding I have left something behind. So my checklists for eclipse trips are as detailed as a mission control flight plan for a shuttle mission. However, making a list is a pain in itself, so, hopefully my list will save you time before your eclipse expedition. I have subdivided my lists into different sub-lists for convenience. Of course, copies of these lists and this book should be packed as a priority!

Critical Items List

Passports, Visas (if required) and a photocopy of your vital Passport pages (just in case); Plane Tickets, Itinerary, Credit Card (check it doesn't expire half way through the holiday!) and appropriate currency; mobile phone (check it will work where you are going, is topped up if using a "pay as you go" tariff and full with the phone numbers of your travelling companions); plus, don't forget your clothes! Plus, a map of the way to your airport or pre-flight hotel is essential!

If doing a lot of tourism, the *Lonely Planet* Guides can be invaluable. If you are doing a lot of walking you may well be tempted to take a pair of stout walking boots that will work well in, e.g., a desert environment, or anywhere where you may turn your ankle. However, break such boots in weeks or months before you embark. Otherwise you will be on the proverbial one way moped to blister city.

A Camera Equipment List

Cameras; camera user manuals; manual shutter release cables (Fig. 9.5); lenses, telescopes and binoculars; tripods (preferably with smooth motion adjustment); solar filters; spare memory cards (or film); laptop interface cables/flash card reader (optional); focussing aids (e.g. a clip on 3× magnifier as shown in Fig. 9.6); tripod thread nuts, bolts, spanners and Allen keys; all telescope to camera bayonet adaptors/eyepiece adaptors; all international mains plug adaptors; all camera battery chargers; spare batteries; every conceivable screwdriver; check you know how to disable each camera's flash (stick black tape over it if necessary!); equipment bag for eclipse day; air blower for removing dust; plastic bags for storing parts in on eclipse day; video cameras; video camera user manuals; video camera

Fig. 9.5. The Canon RS60-E3 remote shutter release. An inexpensive and essential item for "hands free" photography at long focal lengths. Image by the author

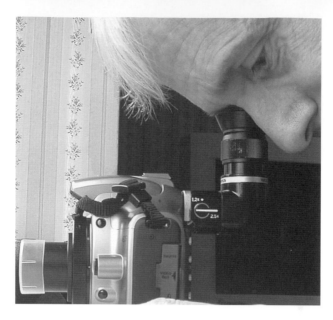

Fig. 9.6. A 2.5× or 3× magnifier attached to the viewfinder can be an invaluable focusing aid during solar eclipses, provided a safe solar filter is used during the partial phases. Image by the author

charger; plenty of spare cassettes, discs or recordable media; clip on microphone for interviews on eclipse day (wind noise on the main mic. can ruin the sound); headphones to check the audio recording quality; rubber bands and sellotape for emergency repairs! You will also need a torch, plus spare batteries, with a red plastic film over the front for seeing your camera controls during totality. (The red film prevents you from irritating nearby observers).

An extra word about Digital SLR camera batteries will not go amiss at this point. After a year or two of use, they often seem to become very pathetic indeed, only lasting half, or a quarter of their original duration. When modern camera batteries are not used to their limit and then fully charged (i.e. when they are charged after small usage) they can go soppy on you. Investing in some brand new batteries, pricey though they are, may save your life at an eclipse site, especially a cold site. When temperatures drop below zero, even modern batteries lifetimes take a hammering. Oh, and another thing: double check whether your tripod head will prevent a quick battery or memory card/video tape insertion in those panic-stricken moments during totality. You have been warned!

Miscellaneous Eclipse Gadgetry List

Watches (with alarms and backlights) accurately synchronised before departure; a GPS receiver to tell you exactly where you are on eclipse day (Fig. 9.7); detailed eclipse track charts and timings for the big day; an MP3 player/Ipod to while away the utterly tedious hours on long plane and bus journeys (plus suitable charger); a big LED or audio (quiet) system to tell you how time is passing during totality. A pocket magnifier is also useful too.

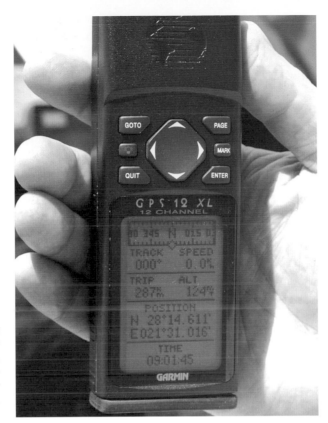

Fig. 9.7. GPS location devices have become very affordable in recent years. This Garmin unit is displaying the location at Jalu for the 2006 Libya eclipse.
Image by the author

Medical Items

These, of course, will vary from individual to individual!

Earplugs for noisy hotels; insect repellent; sunblock cream; sunglasses; wide brimmed sunhat; reading and distance glasses (if used); optrex eye wash for soothing tired, and gritty eyes; chewable aspirin for those horrendous high altitude/ never-ending coach trips; first aid kit; diarrhoea tablets (especially in India!); worn in walking boots/shoes that are suitable for potentially ankle-twisting, rocky and sandy terrain.

Laptops

Eclipse photography is fast becoming a solely digital pursuit and so a laptop can be invaluable for studying your best images; more about this in Chap. 11. However, few people will want to carry their main PC on holiday with them, but a cheaper lightweight model can be invaluable. You can also use it to watch your own choice of DVD movie on the plane flight there, rather than being forced to watch the on-flight movie. That's assuming you have enough battery power! Make sure you pack all the software and electrical leads/adaptors needed for your laptop to talk to your digital camera, video camera and (if required) your flash card/card reader, as well as your favourite image processing software. With a bit of planning in advance you may well be able to devise a way to upload your best eclipse images to your web page while you are on the move to! At the eclipse site, in the partial phases, a cardboard sunshade may be invaluable for keeping the sunlight away from the screen so you can see the LCD display.

Chapter 10

Video, Sketch and Savour the View!

I always find that I am in a dilemma when planning what to do in those brief moments of totality. I am, first and foremost, an astronomical photographer, or, more correctly an astronomical imager. From the first time I became interested in astronomy I wanted to produce photographs. In the 1990s I had to re-learn my dark room and developing skills as the digital era had arrived. Not surprisingly I like to come back from a total solar eclipse with some good digital images. However, I also like to savour the view. Unfortunately, however much savouring I indulge in, I cannot recreate exactly how I felt at the moment of totality; I would need a Star Trek type *Holodeck* to recreate the scene exactly. Weeks after the event, my memories of exactly what I saw in those brief minutes of totality start to merge with previous eclipses, and the precise order that events occur get badly muddled up. In addition, if I spend loads of time fiddling with equipment, when third contact actually arrives I feel like I have badly neglected the real visual spectacle. A compromise has to be made somewhere and, although I still have not sussed the perfect mix of imaging and savouring I understand more about how to approach the experience now, than after my first couple of eclipses.

Sit Back and Enjoy the Spectacle

If you are going to your first eclipse then I would strongly advise you to just sit comfortably, watch the entire event visually, with the naked eye, binoculars and appropriate filters, and then scrounge pictures and videos from friends who were on the same trip. Pictures of the eclipse site, the people, and their equipment bring the whole experience flooding back. A brief checklist for looking at during the minutes before, during, and after totality will be invaluable. It is amazing how easily you forget to look for those critical phenomena in the heat of the moment. If you do use a camera, remember to totally disable the flash during totality. If you do not you will annoy loads of people! Remember to look for the Moon's shadow approaching. It does not have a hard edge, but one horizon should increasingly look darker than the other with a few minutes to go to totality. Make sure you know where Venus, Mercury and other well-placed planets and bright stars are placed with respect to the eclipse and the horizon too (Fig. 10.1). If Venus is at a long, safe distance from the solar disc you may well spot it appearing 10 min before totality. Perhaps most important of all, with less than 2 min to go to second

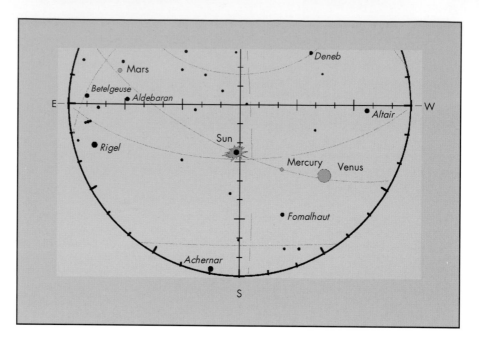

Fig. 10.1. Eclipse chasers like to know where the bright planets and stars will be with respect to the eclipsed Sun at totality. Fortunately NASA publishes an eclipse bulletin for every total eclipse and Fred Espenak and Jay Anderson provide diagrams such as this one. Image: Nasa/Espenak/Anderson

contact DO NOT forget to look for the rippling shadow bands (Chap. 5) on the ground. The lighter the surface you are sitting on, the better for this purpose. A desert of white sand is ideal! You will kick yourself if you don't see the shadow bands and everyone else does; they appear after totality too of course, if conditions are right. Being seated within earshot of experienced fellow travellers can be a great advantage too, as if they spot something unusual you will hear their exclamations. Savour the twilight sky colours and the behaviour of the local wildlife as well; they are all part of the experience.

Sadly, in these days of the SOHO (Solar and Heliospheric Observatory) spacecraft coronograph and automated discovery telescopes, the era when people discovered comets at total solar eclipses is probably resigned to history, or is it? Well, probably! The famous example is that of the aptly named Eclipse Comet of 1948 (C/1948 VI = 1948 XI) which was discovered at the total solar eclipse of 1 November 1948 at magnitude −2. It was next seen again in the morning twilight on 4 November 1948 as a zero magnitude comet with a 20° tail. Despite the microscopic likelihood of this happening during a modern totality a part of me half expects that as second contact takes place a magnificent comet might appear. Well, there is no harm in dreaming . . ..

On every eclipse expedition there are always a few experienced visual observers present, who can accurately sketch what they saw at totality (see Figs. 10.2 and 10.3). This, of course, is the ultimate personal record, providing you are a good artist. It might be thought that trying to sketch the extent and shape of the solar corona during totality would be the ultimate distraction from actually savouring the view. However, most eclipse sketchers that I have seen have simply studied the

Fig. 10.2. Despite the digital imaging revolution the skilled artist can still excel at capturing the atmosphere at totality as the human witness remembered it. Here, the renowned space artist David A. Hardy has used a thin portrait format to convey the high altitude of the eclipse at Jalu in the Libyan Sahara on 29 March 2006, together with a Libyan observer strolling past. Copyright © David A. Hardy. http://www.hardyart.demon.co.uk/.

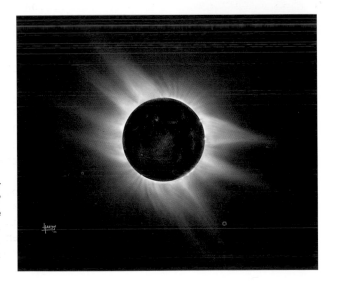

Fig. 10.3. A magnificent colour painting of the 29 March 2006 total solar eclipse by space artist David A. Hardy. Copyright © David A. Hardy. http://www.hardyart. demon.co.uk/.

corona intently during totality, making a few written notes (without looking at the paper) and then, as soon as daylight returns, they have made a sketch in the 10 min that followed. The sort of notes they have written relate to the length and position angle of each coronal streamer (e.g. three solar radii, tapering streamer at three o' clock) and any prominences on the limb, i.e. memory-jogging items that will enable them to complete a sketch immediately after third contact. On arrival home the sketch can be turned into a prettier painting with, perhaps, a check against images taken by friends.

Camcorders

Without doubt, a video camcorder (Fig. 10.4) is the best way of capturing the scene and the mood at an eclipse site. Camcorders can run unattended throughout the critical period of totality and many observers simply leave one running on a wide angle setting throughout totality to capture the horizon colours, verbal exclamations and arrival/departure of the Moon's shadow. Personally, I have found that interviews with friends, before and after totality, explaining their equipment and observing strategy bring back the mood of each eclipse perfectly.

Choosing a camcorder suitable for eclipse trips can be a complex process, not least because of the bewildering number of storage media that seem to change month by month. Critical to the decision will be whether you wish to zoom in on totality with the supplied lens, or purchase additional 2× converters and the like to increase the zoom power, and also whether you can easily edit the video output,

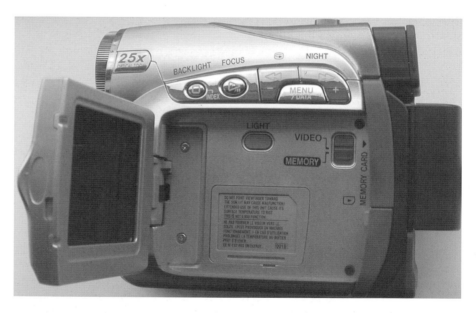

Fig. 10.4. Compact modern camcorders, like this JVC GR-D290EK, feature optical zoom powers of up to 25× which are sufficient for recording reasonable detail at totality without additional 5× or 10× teleconverters. They also record the jubilant sounds at totality and can be used to film entertaining interviews with fellow travellers.

via *Firewire* or High Speed USB interfaces, on your PC. If you are a complete video novice you will need to know that it is only the optical zoom that you are interested in, and not the digital zoom. Many specialist, but affordable, lens accessory manufacturers (e.g. Sigma and Tokina), make video converter lenses of between 2× and 10× power which simply screw into your existing lens. These lenses work well on the camcorder's full zoom setting, but vignette the field badly as you lower the zoom. Nevertheless they can work very well for solar eclipses, even if they can look a bit unwieldy when attached to your camcorder.

The first calculation that most experienced amateur astronomers will attempt to make when purchasing a camcorder for eclipse use is how big the image of the Sun will appear on the chip and how many pixels it will cover. However, performing this calculation means you have to enter the jargon-filled world of image sensor dimensions! In this surreal world you will often find meaningless phrases like "2/3 in." CCD sensor. Sometimes it will say this is a diagonal imaging area measurement. Ha! What utter drivel!! In fact, a typical 2/3 in. CCD sensor has imaging area dimensions of around 6 × 9 mm, way short of 2/3 in.! But this might give you a guide to how big your image sensor really is, and many camcorder sensors are no more than a couple of millimetres across. As a rule of thumb most camcorders have a normal (1× zoom) field of view about 40° wide. So a 10× zoom will reduce this to 4° and 20× to 2°. The Sun and Moon are, of course, only half a degree across, but the full extent of the corona can span many degrees. If your camcorder manual tells you the field of view at various zoom settings then you are laughing, but many don't. However, any camcorder with a zoom of 15× or more will give you a nicely framed corona shot with some structure visible. It should be borne in mind though that typical camcorder still frames will not compete with the quality of a digital SLR or even a high zoom non-SLR digicam image of a solar eclipse. Camcorder electronics are geared for very fast high gain downloads when working in video mode and typical frames from a camcorder video (Fig. 10.5) will be noisy and overexposed when compared to digicam images. There is no substitute for a physically big CCD chip and a long focal length where still frames are concerned. If left on auto exposure a camcorder will, almost always, burn out the coronal detail. Having said that, if you have deep pockets, and can afford a top quality video camera then you can almost get the best of both worlds. At the top end of the market interchangeable lens video cameras, with selectable exposure times, as used by the professionals, will do a great job.

However, if you simply want to lie back and enjoy the eclipse, with your camcorder recording a pretty video, and recording the exclamations of your friends, then a camcorder cannot be beaten. Quite a few of my colleagues own two camcorders and use one to video the Sun and corona while the other just sits a hundred metres away filming the overall wide field scene and the group of people they are with. Alternatively, the second camera can just point at a white card or sheet on the ground waiting for the shadow bands.

Camcorder adverts often boast absurdly large digital zoom values but they are simply software re-sampling tricks to make the image bigger; they do not increase the resolution. A few camcorders offer genuine optical zooms of 20 or 25× which will typically have a field of view a degree or two wide: ideal for close up work on the corona and prominences. The key to successful high resolution eclipse videos is setting the focussing and shutter exposure function to manual and quickly tweaking the exposure as you remove the solar filter at 30 s to second contact, i.e. just before totality starts. Some camcorders do have auto-exposure functions that

Fig. 10.5. This sequence of freeze-frame video stills, shot by the author at the 24 October 1995 total solar eclipse at Fatephur Sikri, India, are typical of what can be recorded at totality with a camcorder.

can react quickly and accurately to the dramatic light levels occurring when the solar filters are removed (massive brightening) and as the first diamond ring slowly fades into totality. However, as a diamond ring and a total solar eclipse are very unusual objects the auto-exposure software might not cope perfectly. Focussing can be even more problematic, as many cameras will simply hunt for a best focus on the black lunar disc as the second contact diamond ring fades from view. Of course, if you are determined to savour the view visually you will not want to fiddle about with the camcorder controls at all during totality, in which case simply leave the focus fixed at the same point as it was with the solar filters on and trust the auto-exposure to cope. An intuitive "feel" for the camera controls in the twilight conditions of totality will be invaluable here and some practice runs on, for example, the Full Moon, at night, in the weeks before, will pay dividends.

You may not be too bothered about high resolution video recording of course, you may just wish for a record of the event and a sound track of the whoops and cries. However if you do want a high magnification video record of the corona and prominences you need to bear in mind that, as the Earth rotates, the Sun and Moon will drift their own diameters in only 2 min of time. Unless you have brought an equatorial mount along, a long duration eclipse may see the entire Sun/Moon drift right out of the field of view, while you are savouring the visual experience. Therefore a lower optical zoom than the maximum (e.g. 15×) might be preferred. Again, the Full Moon at night can be used to practice on.

Just after the October 1995 eclipse in India I was on a coach leaving the site when a fellow traveller suddenly announced, in the most depressed voice imaginable, that he had erased his superb video of totality, less than an hour after he had recorded it. How had this nightmare occurred? Well, he had been rewinding the tape and showing the video to some friends on the coach. He then decided to capture an amusing incident on the video and pressed record, subconsciously thinking the eclipse had passed and therefore he was on the last few minutes of the tape. Tragically, he had rewound to before the brief 45 s of totality and then pressed record. His personal video of the tape was erased for ever. So, whatever media you are using to record totality, always ensure, by removing that tape, DVD, memory card, or whatever, that it is not available for recording after the big event. If using media that is easily removable, stick a label on the tape or mark the DVD with a pen, stating categorically that it is the totality video!

A few other essential checks need carrying out to avoid sheer terror during the critical period! Fresh, fully charged batteries should be loaded into your camcorder 10 min before totality. It is easy to overlook this aspect and then face the horror of hearing your camcorder shutting down during those special few minutes. Whatever recording media you are using should have plenty of time left to fit totality on. It is all too easy to film loads of interviews with colleagues during the partial phases and, in all the excitement, forget that you are low on recording time and battery life. I have seen some camcorders that have an auto-shutdown feature too; in other words if there has been no user activity for 10 min the camera shuts down, whether recording or not! Pre-eclipse dry runs with the camera should iron problems of this nature out.

Aside from the eclipse itself, interviews with colleagues can easily be wrecked if conditions at the eclipse site are windy. The LCD viewfinder might make it look as if all are proceeding smoothly, but when you play the tape back in your hotel room you find that your more quietly spoken friends' comments are trashed by wind noise. The solution to this is to use a clip-on microphone for interviews and bring some lightweight headphones along so you can monitor the audio recording as it is being made.

Solar filters for camcorders are, admittedly, a bit of a problem. Most modern camcorder lenses do have standard filter threads which will take screw-in neutral density filters. However, what is required is a quick-release minimum hassle system where you can whip a solar filter off easily before second contact. In this regard a home made system is almost always best. Provided you are simply looking at an LCD screen when observing the partial phases of an eclipse you can rely on a home-made filter system. If you destroy your camcorder's CCD, that is a shame, but you can replace it. You cannot replace your retina. Modern camcorder CCD chips are far more tolerant of overexposure than in the days of vidicon tubes. For those of you old enough to remember I am sure you will recall the lack of live video footage on the Apollo 12 mission. The camera was accidentally pointed at the Sun and that was that. I have removed my camcorder solar filters a full minute before second contact (admittedly on an ageing camera, which was expendable) and it survived Okay. However, the heat build up inside the camera did cause unfortunate condensation effects after the eclipse as the camera's insides cooled down. All my interviews after third contact at the Zimbabwe eclipse looked like they were made through fog! A sensible compromise is to remove the solar filters

30 s before third contact; it is very handy to have an accurate time source with you and be at the latitude and longitude you have planned for, to achieve this. Modern GPS systems are invaluable in this regard. For my total solar eclipse expeditions I have made camcorder filters from thin cylinders of card, with a film of Mylar or optically dense black polymer strapped across the end. Again, some experimenting weeks before the big day is crucial and it is easy in this case. Also, less opaque filters than normal can be safely used as long as an LCD screen is being used to view the image. However, if an optical viewfinder is available you must not use anything that is less dense than neutral density 5. While the solar corona's brightness can only be approximated by experimentation on a full Moon, the Sun is available every clear day. I have to say that I vastly prefer using the black polymer neutral density filters available from solar filter specialists or shade 12 or 14 "welding goggle glass" for home-made camcorder solar filters, rather than ultra-shiny Mylar film. Unless the gap between Mylar film and camcorder lens is light-tight, various unwelcome reflections can arise when using "bodged" home-made solar filters using crinkly, reflective Mylar.

In theory there is no reason why a webcam cannot be used for "video" recording an eclipse, although they are far better suited to high resolution lunar and planetary imaging. (See my Springer book "Lunar and Planetary Webcam User's Guide"). However, you really need a laptop to store the potential Gigabytes of data from a webcam. All things considered, especially with the high compactness, autofocus/exposure capabilities, audio quality and versatility of modern camcorders, they are far more useful at an eclipse site and are great for interviewing people. For high quality stills, the megapixels of Digital SLRs, see Chap. 11, also win out over webcams.

Camcorder Science – Measuring the Solar Radius

In this era of space telescopes, scientific satellites and space probes it may seem unlikely that amateur astronomers, equipped with small telescopes and camcorders, can carry out crucial scientific experiments. However, when it comes to measuring the photospheric radius of our Sun, amateurs can carry out useful research at the north or south limit edges of total or annular eclipse tracks. The Sun is dazzlingly bright and also heats the Earth's atmosphere through which we view it. This creates turbulence (bad seeing) which, even if all other measurement criteria are met, can lead to huge arcsecond errors of its radius. However, with a solar eclipse the gigantic eclipsing body of the Moon, in the vacuum of space provides a seeing-free (almost) method of measuring the solar radius to, it would seem, a precision of ±0.2 arcsec, using amateur telescopes and camcorders. In terms of actual size, with the solar distance normalised to exactly 1.0 AU (149.598 million km) it currently has an accepted radius of around 959.6 arcsec, implying a radius of around 696,000 km. A variation in solar radius of more than 150 km should therefore be detectable from amateur results, providing a meticulous analysis of the data is carried out. There appears to be evidence that the total solar irradiance, i.e. the Sun's radiation output, is directly related to its radius and therefore precisely measuring the solar radius could be crucial to understanding all the causes of global warming.

Attempts have been made to measure the solar radius without going to total solar eclipses. The NASA-GSFC/Yale/ARC Solar Disc Sextant instrument, an 18 cm aperture telescope carried on a stratospheric balloon, made four flights in 1992, 1994, 1995 and 1996 during which 1.0 AU normalised solar radii of 959.5, 959.5, 959.5 and 959.7 arcsec, respectively, were measured. Also, data from the SOHO satellite has been used to estimate the solar radius, although using a "statistical thermal model correction." SOHO was not specifically designed to measure the precise solar radius.

At the time of writing the role of the amateur eclipse chaser is still crucial to this work but the eclipse chasers involved face a terrible dilemma: they have to travel to the extreme north and south track edges where totality will be a fleeting event, with Baily's beads being the highlight to be video-recorded. Having said this, annular eclipses are just as useful (Fig. 10.6) and you sacrifice far less by being at the north or south limit of an annular eclipse, because you would not see the corona anyway. Solar eclipses occur when the Moon is, by definition, on the ecliptic and so there is virtually no latitudinal (up-down nodding) libration problem to worry about. In addition longitudinal (east–west) librations are minimised when you are looking at the lunar polar regions just covering the solar disc. Hence, the usefulness of solar eclipses and the need to travel to the north and south limits. In practice video recordings concentrating on the north and south polar limbs at high power, are much more valuable than lower resolution imaging of the whole disc. Of course, precise consideration of the US Naval Observatory's Watt's Charts, detailing the mountains and valleys on the lunar limb, needs to be incorporated. The equipment the solar radius measurer needs to take with him or her is: one filtered telescope, typically of around 100 mm aperture; a camcorder; a GPS or UTC time signal inserter that stamps the precise time on the video; and a GPS location device so the observer knows his or her location. The experts in this field are IOTA

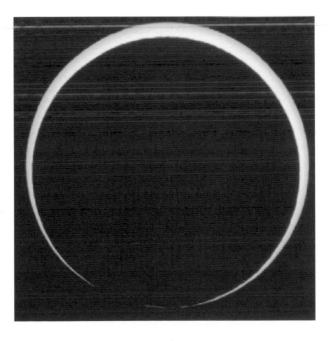

Fig. 10.6. Annular eclipses, at the north and south limits can be used to measure the solar diameter with less sacrifice than for a total solar eclipse (i.e. you will not miss totality). You just need to position numerous observers so they straddle the north and south limits armed with video cameras and accurate timing equipment. The picture shows the 3 October 2005 annular eclipse at the annular/partial threshold, i.e. with the limb mountains breaking up the ring. Image: Damian Peach

(International Occultation Timing Association), led by Dr. David Dunham and you will need to contact them for advice and to be part of a co-ordinated patrol. More information can be found at http://www.iota-es.de/soldiam.html. Typically teams of observers are sent to the north and south limits to straddle the edge of the lunar shadow. IOTA also has the necessary analysis skills and software to interpret your results and allow for the lunar limb Watt's chart correction. IOTA can be found at http://www.occultations.org/. If you are successful you will provide IOTA with a time-stamped video sequence of Baily's Beads, changing split-second by split-second, which they can analyse to a precision of at least 0.1 s of time. Video eclipse observations carried out over the past 25 years do not show the solar radius varying by more than 0.2 arcsec above and below the mean (1 AU) figure of 959.6 arcsec. So, as this is right on the accuracy limit there is no concrete evidence that the Sun has grown or shrunk in radius during this period. Although there is some evidence that the solar radius may have been 0.4 arcsec larger in 1925 and even 1715, this involves comparing historical visual observations with accurate video data and so is very flimsy evidence indeed. Nevertheless, even showing that the solar radius has not varied by more than 150 km in the last 25 years, is an important scientific result. At the Moon's average diameter from the Earth of 31 arcmin, an angle of 0.2 arcsec corresponds to 370 m on the rugged lunar limb. It is not hard to see why the Watts charts lunar limb profile data is crucial to success, or why IOTA will try to straddle the north and south eclipse track edge with observers at 100 m intervals.

Chapter 11

DSLRs and Digital Eclipse Photography

Taking an excellent set of pictures of a total solar eclipse is surely the ambition of many eclipse chasers and amateur astronomers. But how often do the results that you dreamed of obtaining actually materialise? To be honest, this rarely happens without almost obsessive pre-planning, although any decent picture is still a great souvenir of the trip. Total solar eclipses are unique in amateur astronomy: they take place in minutes and you do not get a second chance (until the next trip, which may be years away). In addition, you are usually far from home with, unless you are really dedicated, a distinctly portable set of equipment. Various factors can conspire to thwart the eclipse photographer on the day. Setting your equipment up might be a very rushed affair if you have travelled to the site by coach and hit the inevitable eclipse-site queue. Cloud can thwart the attempt, but so can extreme temperatures (causing equipment failure), wind, sand, tortuous camera angles (with the Sun at the zenith) and even sheer panic. In addition the sheer excitement of the moment can badly affect the photographer's concentration. From a technical viewpoint long lens focus positions can alter drastically as the partial phases proceed and the temperature drops. In addition vibrations from the camera's reflex mirror (and shutter with film cameras) can badly blur the view. If you have not tested your equipment properly you may find that small telescopes have gone out of collimation in transit, removing any chance of getting a sharp image. If you have assigned savouring the visual spectacle a very high priority then less-than-perfect pictures may not bother you too much. However, most experienced eclipse chasers who have seen half a dozen totalities really do want to get a cracking result, so, for them, a technical disaster is a real disappointment. Over the years there have been many articles and books that have been written on the subject of eclipse photography. However, at the 29 March 2006 total solar eclipse, which I observed from the Libyan Sahara, it was obvious that there had been a major revolution in eclipse photography. At total eclipses prior to 2006 digital SLR cameras had been a rarity at the eclipse site. But in 2006 the vast majority of serious eclipse photographers I was travelling with had a digital SLR as their main instrument. In addition, many travellers had brought laptop PCs along so they could download, examine and process their eclipse pictures in the next few days onboard our cruise ship. So many digital images were taken on eclipse day that it was possible for a major Powerpoint presentation to be given in the ship's lecture theatre a few days later, using dozens of excellent results from the trip. It seemed like a new era of digital eclipse photography had dawned and the old days of the

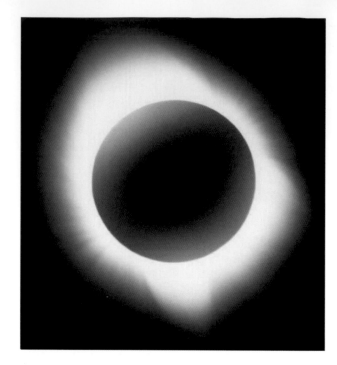

Fig. 11.1. A traditional corona shot (taken by the author using photographic film and no image processing) at the 26 February 1998 Total Solar Eclipse, seen from Knip Bay, Curacao. Celestron C90 f/11 Maksutov (1,000 mm focal length) and 50 ASA Fuji Velvia film; 2 s exposure.

traditional film-based shot, with the corona being a featureless white blob (see my shot in Fig. 11.1) had slipped into history.

The New Digital Era

Digital imaging has some distinct advantages over using film. There is no shutter or film power-winder vibration. Neither is there any risk of film being scratched by particles of sand as it is wound on, or of the film not engaging in the sprocket when loaded. There is also instant confirmation on the LCD screen that a picture has been taken and is safe in the memory. In addition, once the images have been taken they can be examined even on the coach trip back (no waiting for the shopping mall to lose your film or damage it!) and, with powerful image processing routines the maximum detail can be extracted from the images. This latter point is especially valid when imaging the coronal detail which extends over a huge brightness range. So I make no apology for concentrating on digital photography in this chapter.

One thing to bear in mind with digital SLRs, especially when changing lenses, or using a telescope, in a sandy, windy, desert site, is that small pieces of dirt and grit can still easily get inside the camera body. In the days of film a piece of grit could sit in the film gate and score a line along every frame as you wound the film on. This will not happen with a digital detector of course, but you really do not want a small particle of sand sitting on your expensive CCD or CMOS chip. So, if you are headed to a dirty windblown site, you may wish to consider assembling your camera and lens/telescope arrangement in your hotel room before venturing to the site. Conscious of the dust problem, in 2006 Canon introduced their ambitiously worded "EOS Integrated Cleaning System" with their 400D model. Most of the innovations involved making the interior of the camera body from materials less likely to generate dust, but anti-static components were introduced on their sensors too and an anti-alias filter which vibrates to shake off dust was incorporated. For stubborn dust, a "Dust Delete Data" software feature was introduced to process out permanent dust problems.

Digital SLRs can be used in two configurations for high resolution eclipse photography, namely, using telephoto camera lenses with the appropriate bayonet adaptor, or, using an adaptor to fit into the throat of a telescope drawtube (Fig. 11.2); in the latter case the telescope lens (or mirror) then replaces the camera's own lens.

Non-SLR digicams, i.e. digital cameras without a "through-the-lens" optical viewfinder, and without the option of interchangeable lenses, can also be used for eclipse photography. Such cameras rarely come equipped with a built-in powerful enough zoom lens for high resolution work, but they do have zero vibration as there is no reflex mirror at all. The main issue with such cameras is obtaining a sufficient focal length to take high resolution images. It might be thought that there is no way of attaching a fixed lens digicam to a small telescope. However, the US company *Scopetronix* (http://www.scopetronix.com) has, for a number of years, made a whole range of useful adaptors and eyepieces specifically designed to interface fixed lens digicams to telescopes. The approach in this case is to put the camera where the human eye would be, i.e. at the eyepiece. Of course the human eye can get really close to an eyepiece, close enough for the eye's own pupil to be optimally positioned in relation to the "exit pupil" of the eyepiece. If you try getting a standard digicam that close you will, quite possibly, clang the camera lens into the eyepiece and the resulting image will be highly vignetted too, i.e. it will look like you are peering down a drainpipe. The *Scopetronix* adaptors make this a lot easier by using a special wide field eyepiece with a huge, flat, eye lens and by marketing adaptors that rigidly grab the digicam and hold it safely in place with the camera lens almost touching the eyepiece lens (Fig. 11.3). To minimise vignetting the camera needs to be on a high zoom, much as with the camcorder video extender lenses. So, fixed lens digicams can be used for high resolution

Fig. 11.2. With the lens removed a standard digital (or photographic) SLR camera can be attached to a small telescope using the appropriate bayonet to 31.7 mm adaptor. Image by the author

Fig. 11.3. A Scopetronix Maxview eyepiece and digi-cam adaptor (this one for a Nikon Coolpix 5700) used to join a fixed lens digicam to a telescope. Image by the author

eclipse photography, but ever since the advent of the Canon Digital Rebel/300D, DSLRs have been affordably priced and are the number one option.

Digital SLRs

The first affordable DSLRs all had CMOS or CCD detectors that were roughly 2/3 the size of the old, "35 mm" standard film frame of 36 × 24 mm. Specifically, in the Canon Digital Rebel/300D cameras, the CMOS detector is 22.7 mm × 15.1 mm (with 3088 × 2056 pixels). This detector size is usually classed as an "APS-C" size sensor as it is similar to the film based Advanced Photographic System's "C" or Classic format of 25.1 × 16.7 mm which, introduced in 1996, preceded the digital era by a few years. Thus, with the smaller chip (or film) the field of view when compared to the same lens on a "35 mm format" film camera would be about 1.5× less. Put another way, the resulting field of view was like using a lens 1.5× longer. A 300 mm lens on a 2/3 size sensor will have the field of view of a 480 mm lens used on a 36 × 24 mm film frame. However, at the time of writing (2007) full-frame "35 mm format" sensors with 36 × 24 mm chips are now also becoming affordable. Needless to say, things are not quite as simple as this. Are they ever? Individual manufacturers' sensor sizes differ considerably and there is a distinct category of sensor midway between APS-C and full frame 36 × 24 mm sensors. Canon's 1D Mk II sensor has dimensions of 28.7 × 19.1 mm and so is often referred to as an "APS-H" (H = high definition) sensor, even though an APS-H film frame is actually defined as 30.2 × 16.7 mm. Table 11.1 details a variety of Canon and Nikon DSLRs with various sizes of sensors and their critical resolution (arcseconds per pixel) and field of view values when used with a typical eclipse focal length of 400 mm. To calculate these values for any device you only need to use two formulae.

Firstly, arcseconds/pixel = 206 × pixel size in micrometres/focal length in millimetres.

Secondly, for medium/telephoto lenses:

Field of view = arctan (chip dimension (mm)/lens focal length (mm)).

Table 11.1. A selection of digital SLRs available new, or used, in 2007

Brand/model	Pixel size/no. (µm/Mpixel)	Detector size (mm)	400 mm s/pix	400 mm degrees	Intro date
Canon 300D/ Rebel	7.4µ × 6.3M	22.7 × 15.1	3.8	3.2 × 2.2	8/03
Canon 350D/ Rebel XT	6.4µ × 8.0M	22.2 × 14.8	3.3	3.2 × 2.1	2/05
Canon 400D/ Rebel Xti	5.7µ × 10.1M	22.2 × 14.8	2.9	3.2 × 2.1	8/06
Canon D30	10.5µ × 3.1M	22.7 × 15.1	5.4	3.2 × 2.2	5/00
Canon D60	7.4µ × 6.3M	22.7 × 15.1	3.8	3.2 × 2.2	2/02
Canon 30D	6.4µ × 8.4M	22.5 × 15.0	3.3	3.2 × 2.1	2/06
Canon 20D	6.4µ × 8.2M	22.5 × 15.0	3.3	3.2 × 2.1	8/04
Canon 20Da (Astro)[a]	6.4µ × 8.2M	22.5 × 15.0	3.3	3.2 × 2.1	1/05
Canon 10D	7.4µ × 6.3M	22.7 × 15.1	3.8	3.2 × 2.2	2/03
Canon 5D	8.2µ × 12.7M	35.8 × 23.9	4.2	5.1 × 3.4	8/05
Canon 1D	11.0µ × 4.2M	27.0 × 17.8	5.7	3.9 × 2.5	11/01
Canon 1D Mk II/IIN[b]	8.2µ × 8.2M	28.7 × 19.1	4.2	4.1 × 2.7	06/04
Canon 1Ds	8.8µ × 11.1M	35.8 × 23.8	4.5	5.1 × 3.4	12/02
Canon 1Ds Mk II	7.2µ × 16.7M	36.0 × 24.0	3.7	5.1 × 3.4	10/05
Nikon D50	7.9µ × 6.0M	23.7 × 15.5	4.1	3.4 × 2.2	5/05
Nikon D70	7.9µ × 6.0M	23.7 × 15.6	4.1	3.4 × 2.2	1/04
Nikon D70s	7.9µ × 6.0M	23.7 × 15.5	4.1	3.4 × 2.2	5/05
Nikon D80	6.1µ × 10.0M	23.6 × 15.8	3.1	3.4 × 2.3	9/06
Nikon D100	7.9µ × 6.1M	23.7 × 15.6	4.1	3.4 × 2.2	2/02
Nikon D200	6.1µ × 10.0M	23.6 × 15.8	3.1	3.4 × 2.3	1/05
Nikon D2Hs	9.6µ × 4.0M	23.7 × 15.5	4.9	3.4 × 2.2	2/05
Nikon D2X[c]	5.5µ × 12.2M	23.7 × 15.7	2.8	3.4 × 2.2	9/04
Nikon D2Xs	5.5µ × 12.2M	23.7 × 15.7	2.8	3.4 × 2.2	6/06

Needless to say, this table will date rapidly as the years go by! Nevertheless it gives an indication of the cameras being used by eclipse chasers at the 29 March 2006 eclipse and those planning for the 1 August 2008 eclipse. In my experience, Canon DSLRs are the most popular cameras used by eclipse chasers, followed by Nikon DSLRs; hence I have biased the table towards this trend. The pixel size/number of column details the size in microns (thousandths of a mm) of each pixel, followed by the number of megapixels on the sensor. Detector size, in millimetres, is self-explanatory. The next two columns are, specifically, for a lens focal length of 400 mm, a useful general purpose focal length for eclipse photography. They detail the arcseconds covered by each pixel (1 arcsec = 1/3,600 of a degree) and the field of view in degrees covered by the whole sensor with a 400 mm lens. An excellent online source for camera reviews can be found at: http://www.dpreview.com/reviews/

Some caution is needed when looking for the same cameras in the USA and UK. The same cameras can have different names, e.g. the Canon 300D in the UK was named the Canon Digital Rebel in the USA

[a]The Canon 20Da was a Canon 20D with more red sensitivity for Deep Sky astrophotography. It was discontinued in 2006

[b]The differences between the Canon 1D MkII and MkIIN are very subtle, namely a wider viewing angle LCD, improved buffering and the ability to write different formats simultaneously to SD and CF cards

[c]The Nikon D2X and D2Xs models feature CMOS detectors, whereas all other Nikons use CCDs. At the time of writing all Canon DSLRs in production use CMOS detectors

Focal Lengths, Fields of View and Resolution

Essentially there are two types of long lens eclipse photography: ultra-short high resolution prominence shots, and wider field shots that reveal the delicate and intricate nature of the corona. Only the best images, processed with much care and expertise show the full range of detail visible to the human eye (Fig. 11.4).

If one goes back over the past few decades and looks at the strategies adopted by eclipse photographers in the film era, the good old 1,000 mm lens was the hallmark of the prominence photographer, i.e. the person aiming at freezing the fine, 1 or 2 arcsec detail, on the second contact and third contact limbs, the diamond ring, and even Baily's Beads. In the 1970s and 1980s the Celestron C90 Maksutov telescope, a 90 mm aperture, 1,000 mm focal length f/11 system, was the "de rigeur" item in the eclipse chaser's carry-on baggage. This compact, low mass, system was (and still is) a very useful package to take on eclipse trips. The favorite film of that era was the slow Kodachrome 64 which was one of the few films of that time to have almost invisible grain. Of course, as film technology improved throughout the 1990s even ISO 200 and 400 films had acceptable grain. Unfortunately, a 64 ISO film at f/11 had an optimum prominence photography exposure range right in the middle of the old film camera's shutter vibration territory. Unless you had a rock solid system, exposures of 1/30 or 1/60 with a 1,000 mm lens led to many double-images of prominences. At 1,000 mm the half degree disc of the average Moon, silhouetted against the bright corona, spans $1000 \times$ tangent $(0.5°) = 8.7$ mm, on the film or CCD/CMOS detector. In practice, as we have seen, the Moon can be quite a bit larger than this, but let's not quibble over a few arcminutes. We can also easily calculate that a 36×24 mm "film-sized" detector will cover $2.1 \times 1.4°$ at 1,000 mm focal length (e.g. $1000 \times$ tangent $(2.1°) = 36.7$ mm) and a 23×15 mm detector will

Fig. 11.4. The full range of subtle coronal detail, 1 s before second contact, is captured in this superb 64 image composite taken from a 1,330 m high peak near Göreme, Cappadocia, Turkey on 29 March 2006 by Hana Druckmüllerová. Expertly processed by Hana and her father Miloslav Druckmüller. 500 mm f/8 Maksutov-Cassegrain Russian mirror lens (MC 3M-5CA) on GS equatorial mount. Canon EOS 1D Mark II (ISO 100). Exposures from 1/1,000s - 8s. © 2006 Hana Druckmüllerová/Miloslav Druckmüller.

cover a smaller $1.3 \times 0.9°$ field. So a 1,000 mm lens, if used with a full-size or a 2/3 size detector, will nicely fit the Sun/Moon and prominences on the chip, although the full extent of the outer corona is too large for this focal length. In the days of film, film resolution was quoted in terms of high contrast line-pairs per millimetre that could be resolved by the film grain. Typically, the slowest, finest grain films could resolve 200 line pairs per millimetre or two lines that are 5 μm apart. According to a law called the Nyquist Sampling Theorem you would need pixels half as finely spaced as this (i.e. 2.5 μm apart) to match this film resolution. In many modern digital SLRs the pixels can be three times this size/separation, i.e. 7 or 8 μm across. Eight micrometres equates to 125 pixels per linear millimetre and, by Nyquist theory, the ability to split roughly 60 line pairs per millimetre. So it might seem, on the face of it, that a larger focal length would be needed for resolving fine detail at eclipses than in the days of film; perhaps a 2,000 mm lens is required? In fact, the reverse is true, because although the old film resolution charts are correct the ability to resolve high contrast black and white lines is easy when compared to resolving subtle, low contrast and overexposed detail, i.e. the sort of detail that astronomers try to capture. In addition, pixels have a fixed size whatever amount of light you throw at them, whereas film grains get big and clumpy when overexposed. Also, you can increase the ISO rating on your DSLR to 200, 400, 800 or 1,600 ISO and the pixels still stay the same size, even if the pixel to pixel noise increases. With film the grains get bigger as you go faster. In addition, image processing sharpening techniques, applied to digital images, can bring out fine details that are not immediately obvious in the original image. Taking all these factors into consideration, and looking at all the results from the 2006 total solar eclipse, a Digital SLR equipped with a 400 or 500 mm lens can easily match a 1,000 mm lens using even slow film. Set the ISO rating to 400 for a faster exposure, and with the benefit of zero shutter and film winder vibration the results can be stunning. In general bigger pixels and bigger detectors are better, because bigger pixels are simply less noisy. Obviously if the pixels in a particular full frame sensor are really big, e.g. larger than 10 μm, you may want to move to a longer focal length lens; the larger sensors that tend to go with larger pixels will still give you a decent field of view. Another consideration, regarding ISO settings on digital SLRs, is that increasing the ISO rating tends to decrease the dynamic range of the detector, and for capturing the corona, a large range of brightness values need to be tolerated. At the ISO 100 setting a typical DSLR detector can capture 11 or 12 photographic stops (a brightness range of a few thousand) but at ISO 1000 only nine or ten stops might be tolerated (a brightness range of 500–1,000 times).

There are two other big advantages of coming down in focal length too. Firstly, the eclipsed Sun is far easier to relocate should you knock your tripod in a crisis at totality. It is remarkable just how tricky it is to position a half degree Sun in a 1° frame when your neck is cricked, the Sun is near the zenith, the tripod leg is sinking into the sand and you've brought a flimsy tripod to save weight. Actually, I will re-phrase that, it is not remarkable, it is damn obvious! At 400 mm you have over six times more area to play with. The second advantage is that much of the outer corona can fit into the field of a 400 mm lens. With the Canon 300D/Digital Rebel 2/3 chip size of 22.7×15.1 mm the 400 mm field will be $3.2 \times 2.2°$. With the Canon 5D full-frame chip size of 35.8×23.9 mm the field will be $5.1 \times 3.4°$. In either case the field is very appropriate for capturing the outer solar corona and equatorial streamers. A 3° field is equal to 12 solar radii, or five solar radii beyond each solar limb. So with a 400 mm lens on a DSLR you can capture fine prominence and outer

Fig. 11.5. A Canon 1D digital SLR equipped with Canon's excellent EF 100–400 mm f/4.5–5.6L lens which also has an image stabiliser option. A very suitable hand-held option for eclipse chasers on-board ship for whom a heavier lens is just too unwieldy and image stabilisation a must-have option. Image: Damian Peach

coronal detail without changing lenses. Okay, changing lenses might not sound like a big deal, but ask anyone who has tried doing it during the heat of totality: they never try it again! Lenses of around 400 mm focal length come in a variety of *f*-ratios and prices and short focus refractors can often be employed for the same purpose and at a fraction of the cost of the very best lenses. Another piece of technical wizardry to contemplate, especially if you are observing an eclipse from onboard ship, is the growing range of image-stabilised zoom, or fixed telephoto lenses available. To quote a specific example the Canon EF 100–400 mm f/4.5–5.6L IS USM lens (Fig. 11.5) is a real beauty at around $1600 in 2007. Okay, it is a lot of money, but that lens really excels at short exposure handheld eclipse or full/gibbous/half Moon photography. If you are on the heaving deck of a ship you could not choose a better lens if you have a Canon camera body. With the fairly typical 8 micron pixel size found in many DSLR sensors a 400 mm focal length will give an image scale of just over 4 arcsec/pixel, fine enough to resolve details in quite modest prominences, and with roughly 450 pixels covering the mean solar/lunar diameter.

The Best Camera Lenses

When choosing a telephoto lens for eclipse photography there are various considerations to bear in mind. If you are only interested in short exposure Baily's Bead/Diamond Ring/Prominence work, e.g. exposures faster than 1/500, a completely handheld strategy can be the best, especially if you are onboard ship and have a high quality image stabilised lens (Fig. 11.6). Those observers with highly portable, but flimsy, tripods will probably get better results with such photography, by selecting their exposure times below 1/500 or, preferably, 1/1,000 thereby eliminating shake (and, for some, SLR reflex mirror vibration problems). However, if you are wanting to capture the outer corona the longer exposures required may well necessitate a sturdy tripod, and image stabilisation may not be able to cope with the handheld approach, unless you have a very expensive long

Fig. 11.6. The white collar between this Canon 20D body and the 500 mm EF f/4 lens is a 1.4× extender which converts the lens to 700 mm focal length. Also note the third slider down on the main lens body which turns the image stabiliser on; a useful option if you are observing an eclipse from onboard ship. Image: Jamie Cooper

focal length and fast *f*-ratio lens. If you are forced to observe handheld from a ship's deck then make sure you take as many exposures as possible to maximise your chances of some being sharp. Also, many eclipse photographers fancy a go at wildlife photography too. On the Zimbabwe eclipse in 2001 quite a few astronomers used the same image stabilised handheld lens for the eclipse as for the trip to the game reserve and photography of the local warthogs! With camera lenses you pay the serious money when aspheric super colour corrected lenses are used and/or when the piece of glass at the business end is more than roughly 70 mm in diameter. Fortunately fast *f*-ratios are not necessary for land-based eclipse photography as the prominences, diamond ring and Baily's beads are very bright and outer corona shots can be taken from a tripod. We are not trying to capture a cheetah in twilight running after its prey. So the good old 1,000 mm f/11 Celestron C90 Maksutov has more than enough clear aperture for the job. But, I say again, you may well want a faster lens when taking photographs from onboard a ship, unless the thought of hand holding a heavy lens puts you off. Some mouth-watering lenses are available for those with deep pockets though. Canon's EF 400 mm f/2.8L IS USM will set you back a cool $6500 or more and their 500 mm EF f/4L (Fig. 11.7) is not much less at around $5,500! All the "L" series lenses in Canon's range offer exceptional quality colour-corrected optics. If you cannot afford the excellent

Fig. 11.7. Canon's 500 mm EF f/4L lens is a very pricey piece of kit but its fast aperture can quarter the exposure times used with a 500 mm f/8 mirror lens. Coupled to a modern Canon digital SLR it is also a superb and intelligent system for wildlife photography if your eclipse holiday includes some Safari excursions. Image: Jamie Cooper

$1600 Canon EF 100–400 mm f/4.5–5.6L IS USM lens, the fixed focal length 400 mm f/5.6 non-stabilised equivalent can be acquired for under $1100. If you already have a 200 or 300 mm lens, and want a bit more resolution, you may want to investigate purchasing a tele-extender. The Canon range offers 1.4× and 2× extenders for many of their compatible lenses. The extender interfaces between the camera body and the lens bayonet mount for an outlay of a few hundred dollars. Of course independent lens manufacturers like Sigma & Tamron offer a cheaper solution to the premium quality lenses offered by the likes of Canon and Nikon, although the colour performance at the edge of the field will not be as good as with the most expensive brands. In practice the difference in resolution between a 400 mm lens and a 500 mm lens, even with a 1.4× extender added, may not be all that significant (see Fig. 11.8), plus, the pixel size needs to be taken into account to. High resolution is absolutely critical for resolving fine prominence details, but somewhat less so in wide field shots capturing the extreme edges of the visible corona. A shot showing some modest prominence/chromosphere detail, by the author is pictured in Fig. 11.9. Some extraordinary images showing ultrafine prominence and chromospheric detail, image processed by Miloslav Druckmüller are shown in Figs. 11.10–11.14.

In case the reader has become a bit demoralised at the prices I am quoting here, do not be! Expensive lenses are not essential for eclipse photography. Russian lenses are especially good value for money, especially if you have a bayonet to M42 (Pentax) thread adaptor for your camera. The Zenit MC 3M-5CA 500 mm focal length f/8 Maksutov-Cassegrain mirror lens can be acquired for around $160 in 2007 and is a proven performer at past eclipses. Some caution is necessary when purchasing 42 mm thread adaptors as astronomical ones (T adaptors) have a thread pitch of 0.75 mm and photographic ones (Pentax

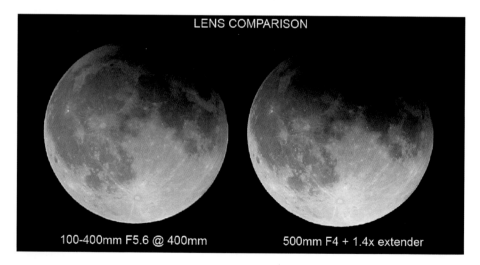

Fig. 11.8. In practice, for bright objects, like a Full Moon, partially eclipsed (which has a similar brightness to the solar corona) there may be little difference between a 400 mm lens at f/5.6 (100–400 mm zoom) and a 700 mm lens at f/5.6 (500 mm f/4 × 1.4 extender). Of course detail in small solar prominences would be more of a test. Canon digital SLR images of the 7 September 2006 partial lunar eclipse made at the same time by Jamie Cooper (700 mm) and Damian Peach (400 mm) and compared by Jamie.

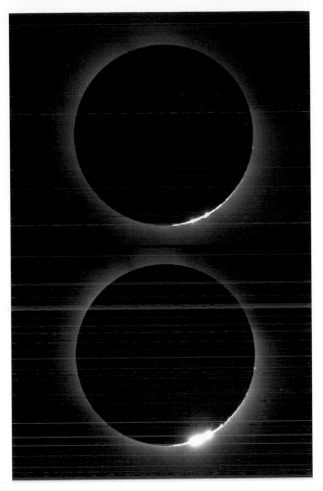

Fig. 11.9. Two images by the author showing the third contact emergence of the double Diamond Ring at the 29 March 2006 total solar eclipse. Celestron C90 (90 mm aperture, 1,000 mm focal length Maksutov Cassegrain) and Canon 300D digital SLR. Both exposures 1/500 at ISO 400. 10:30:41 (*upper*) and 10:30:43 (*lower*) UT.

type) have a 1 mm thread pitch. Thus many astronomers seem to own combinations of astronomical and photographic threaded equipment which jam tightly after one revolution when the threads start to cross. In addition, adding an adaptor can cause difficulty in achieving focus at infinity with some lenses. Nevertheless, Russian Zenit lenses are still a bargain, even if the trend towards DSLRs and, more importantly, modern bayonet fittings, have complicated things somewhat.

Refractors and Maksutovs

Then we come to the option of using short focus astronomical refractors for eclipse photography. Such refractors have a fixed aperture and have to be manually focused; there is no electrical control from camera to lens as with a dedicated camera lens. Neither is there any image stabilisation option. However, many amateur astronomers and eclipse chasers can use these small refractors for visual observing

Fig. 11.10. A superb high resolution image of the chromosphere and prominences at second contact at the 21 June 2001 Total Solar Eclipse, as observed from Ngunza, near Sumbe, Angola, by a team from Úpice Observatory (Czech Republic). A 100 mm aperture, 1,875 mm focal length Mertz Siderostat by VOD Turnov was used, along with a Pentax 6 × 7 medium format camera back and Kodak Ektachrome 100S film. This is the result of a nine image composite, scanned at 2.6 arcsec/pixel, and expertly processed by Miloslav Druckmüller. Copyright © 2001 Úpice Observatory, © 2003 Miloslav Druckmüller.

Fig. 11.11. A second or two before the image shown in Fig. 11.10, the same team captured the last glimpse of the blinding photosphere disappearing at second contact. Same details as for the previous figure. Copyright © 2001 Úpice Observatory, © 2003 Miloslav Druckmüller.

Fig. 11.12. This awesome image of the 8 April 2005 hybrid eclipse is composed of 22 images taken by Miloslav Druckmüller (mainly outer corona) and 9 images taken by Peter Aniol (mainly inner corona) aboard the ship MV Discovery in the South Pacific Ocean. Miloslav's Canon EOS 1D (11.5 micron pixels) and EF 100–400 mm lens (used at 400 mm, f/5.6) gave an image scale of 5.9 arc-sec/pixel. Peter's Canon EOS 20D (6.4 micron pixels) and Canon EF 600 mm f/4 lens (at f/5.6) gave a finer image scale of 2.2 arcsec/pixel. Exposures ranged from 1/125 to 1/4,000 s. (Baily's beads). Images taken through thin cirrus cloud. Copyright © 2005 Miloslav Druckmüller, Peter Aniol.

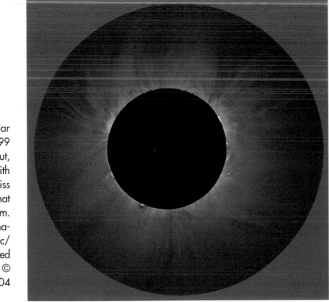

Fig. 11.13. The total solar eclipse of 11 August 1999 photographed from Harput, Turkey by Vojtech Rušin with 200 mm aperture f/15 Zeiss optics and 18 × 24 cm format Kodak PPF 6431 slide film. This is the result of three images scanned at 2.7 arcsec/pixel and expertly processed by Miloslav Druckmüller. © 1999 Vojtech Rušin, © 2004 Miloslav Druckmüller.

Fig. 11.14. The Total Solar Eclipse of 11 August 1999 photographed from 2 km SSE of Németkér village, Hungary by Miloslav Druckmüller. An MTO 1000a Maksutov-Cassegrain was used (focal length 1,084 mm f/10.4) and a Soligor SR 300MD camera with Fujicolor Superia 800 film. Images scanned at 4.8 arcsec/pixel. This image is the result of 22 exposures, combined and processed by Miloslav Druckmüller some five years later. © 2004 Miloslav Druckmüller.

and deep sky photography back home so they can be more versatile. In terms of a cost comparison, prices are similar for the sort of apertures and focal lengths we are considering, although refractors do *not* come in the ultra-fast *f*-ratio of 2.8 that the really pricey lenses by Canon and Nikon offer for sports photographers. But small, high quality refractors like the 60 mm aperture, 355 mm focal length, (f/5.6) Takahashi FS60c offer a budget entry point (around $500) and 500 mm focal length refractors of around 75 mm aperture (f/6–f/7) retail at roughly $1300. Two examples of this latter category, frequently seen at eclipse sites, are the Pentax 75 mm SDHF 500 mm f/6.7 refractor, the Televue 76 mm aperture f/6.3 model (480 mm focal length) and 70 mm f/6.8 TeleVue Pronto/Ranger. These are just three examples, as there has been a virtual explosion in small, high quality refractors available in recent years. The manufacturer Vixen has a high reputation for building quality refractors and its ED 80Sf, 80 mm f/7.5 model is worth a look, as is Celestron's 80 mm f/6.25 (500 mm focal length) Onyx 80EDF. The most important consideration here may well be unrelated to the manufacturer or even to value for money; it is, will the refractor fit in your suitcase? The standard suitcase will let you squeeze a 600 mm focal length refractor diagonally across the case, especially if the dewcap is retractable. But beyond 700 mm it can be a close call! Make sure you have researched this aspect, specifically tube length, before making a purchase. A brand new refractor is nicely cushioned when lying in a case full of soft clothing (clean or dirty), but far more vulnerable when carried separately.

The Meade ETX range of Maksutovs, especially the optical tube assemblies (i.e. with the heavy drive base removed) are quite popular at eclipses and far more compact than a refractor of the same aperture. There is, potentially, one huge advantage of using small aperture, high optical quality telescopes, rather than a lens, for eclipse trips: you can use them visually, as well as photographically. In the heat of the panic-stricken few minutes of totality it is great to have a system where you can switch between camera and eyepiece quickly. You cannot do this with a camera lens, but if you have a telescope, with a suitable flip-mirror unit, you can switch between camera and eyepiece with the turn of a knob. I first found this out at the Zimbabwe total solar eclipse when I brought along a 125 mm aperture Meade ETX Maksutov optical tube assembly. (I left the heavy and unreliable drive base back at home.) That ETX had a long 1,900 mm focal length (f/15) so only inner corona/prominence photography was being considered. The ETX design, similar to the design of the legendary Questar telescopes, allows light to go via a mirror to an eyepiece, or, straight through to a camera. Obviously the eyepiece view is mirror imaged. Thus you can enjoy the best of both worlds. However, with any arrangement using flip-mirrors you will almost certainly have a different focus position between camera and eyepiece. There are two tricks to learn here.

Firstly, focus the telescope for the camera, not for the eyepiece, just before totality (with the filters safely in place). Secondly, well before the eclipse trip make/bodge an adaptor to space the eyepiece at the right position so that camera and eyepiece are in focus at the same time. I say focus for the camera because the eye can compensate for the eyepiece view being a bit out of focus, but the camera's detector cannot. Of course, I have been describing my own ETX in the last few sentences but, in general, short focus refractors do not come with flip mirrors. So you have to acquire a flip-mirror, with a wide 50 mm throat (to avoid vignetting) that is compatible with your telescope. Specialist dealers can provide you with such accessories. Of course, you may well choose a different option, e.g. having a telescope for the visual view, and a second telescope, or lens for the photography. Or, you may prefer binoculars for the visual view and a standard camera lens for the eclipse photography. It is up to you. However, I would still maintain that a focal length of 400 mm or so is the most versatile eclipse photography system. The field is wide enough to easily find the Sun, even on a basic tripod and the field of view/resolution allows outer corona and prominence photography with just one instrument. One point to note here is that with the smaller ETX models annoying lens flares can occur when imaging the brilliant diamond ring phases. Light can sneak past the baffle tubes and flare across the eclipsed Sun. This problem does not seem to occur with the superior baffling around the secondary mirror on the larger ETX 125 models.

Capturing the Corona

Experienced eclipse chasers will be well aware that there is no fixed set of exposures for the range of phenomena visible at a solar eclipse. Even when you know the specific shot you are going for, and the lens f-ratio, and the ISO setting, different eclipses require different exposures. Sometimes the atmospheric transparency is poor and sometimes it is excellent; the height of the Sun in the sky may well be a factor here. The relative size of the Sun and Moon is also important. Even in mid-totality, if the solar disc is almost as large as the lunar disc the brightest parts of the corona, just above the chromosphere will be visible and considerably brighten the sky just around the black lunar silhouette. For all these reasons the eclipse photographer will want to "bracket" his or her exposures; in other words a single exposure of, say, 1/250 is not a safe strategy, try two stops either side, i.e. 1/60, 1/125, 1/250, 1/500 and 1/1,000. This means more work during the critical totality phase, but with practice, and familiarity with camera controls in the dark, you should be able to change exposures as second nature. Many digital SLRs incorporate an auto-bracketing function in certain modes which does this for you.

For recording the most intricate detail in the entire solar corona a huge range of exposures is essential. The human eye has a remarkable ability to see fine details in the inner and outer corona despite the huge difference in brightness between them. Capturing this enormous dynamic range on film, or digitally, has always been a challenge. The faintest outer parts of the visible solar corona, where it fades into the twilight eclipse sky, are more then 10,000 times fainter than the relatively blinding inner corona near to the lunar limb. Capturing the dynamic range is not the only issue either; displaying it has always been a problem too. A projected slide or a glowing PC screen makes the corona look far more lifelike. The coronal structure is quite delicate and so benefits from high contrast processing. However, if you boost contrast you lose dynamic range. We need to enhance delicate structure but fit all the brightness levels into the range of the negative, slide or 8 bit PC screen; a tricky challenge. Nevertheless, the Czech mathematician and master of coronal processing, Prof. Miloslav Druckmüller, has risen to this challenge in recent years, and is the undisputed master of the art. Just study the details in Figs. 11.15–11.19 for proof. In 2002 Druckmüller started his MMV (Mathematical Methods of Visualisation of Solar Corona) project, with the aim of developing mathematical methods to process his and other eclipse chasers' images of the solar corona more effectively. This project placed the emphasis on precise image registration and

Fig. 11.15. This awesome inner and mid-coronal image of the 2006 total solar eclipse really shows the power of combining many different images using custom software when carried out by the digital eclipse magician, Miloslav Druckmüller. The images were taken by Miloslav and by Peter Aniol from Libya and have been aligned to recreate the position of the Moon 15 s after second contact. Two Canon EOS 5D II digital SLR camera bodies were used. Two high-quality apochromat refractors, of the type used by many amateur astronomers, were used for the imaging. Firstly, a Takahashi 100 mm aperture ED f/8.2 refractor combined with a 2× Canon teleconverter, giving 1,640 mm focal length; secondly, a TMB 102 mm aperture f/6 refractor plus Baader FFC 2× converter giving 1,240 mm focal length. With the 8.2 micron pixels of the Canon 5D this translates to 1.0 arcsec/pixel for the 1,640 mm focal length and 1.4 arcsec/pixel for the 1,240 mm focal length. Sixty images with the longer focal length instrument, plus 23 with the shorter instrument were combined for this spectacular result. Exposures ranging from 1/1,000 to 4 s were used. As can be seen, the longest exposures just captured the dark and ghostly lunar surface illuminated by Earthshine. © 2006 Miloslav Druckmüller, Peter Aniol.

Fig. 11.16. This is another image of the 2001 Total Solar Eclipse by the Úpice Observatory team who observed from Angola (see Fig. 11.10 for equipment details). This wider field corona shot was the result of combining eight images scanned at 4.4 arcsec/pixel. Image processing by Miloslav Druckmüller. Copyright © 2001 Úpice Observatory, © 2003 Miloslav Druckmüller.

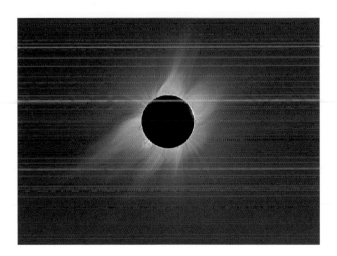

Fig. 11.17. Two of the biggest names in eclipse-chasing combined forces to produce this excellent image of the total phase of the 8 April 2005 hybrid eclipse. NASA's Fred Espenak (Mr. Eclipse at http://www.mreclipse.com) took the seven individual exposures (1/2,000 to 1/15) from the deck of the MV Galapagos Legend in the South Pacific. He used a Canon EOS 1DS (Mk II) digital SLR and a 500 mm EF f/4 Canon image stabilised lens (see Figs. 9.2, 11.5 and 11.6 for Jamie Cooper's identical lens). Remarkably, Fred was coping with a rocking ship during the eclipse, but managed, with the image stabilised lens, to keep the eclipsed Sun in the frame for the 32 s of totality and even cope with exposures as long as 1/15 s. (System field of view and resolution: 3.6 arcsec/8.8 micron pixel and 4.1 × 2.7° field. Miloslav Druckmüller processed the images to resemble the view in a small telescope. © 2005 Miloslav Druckmüller, Fred Espenak.

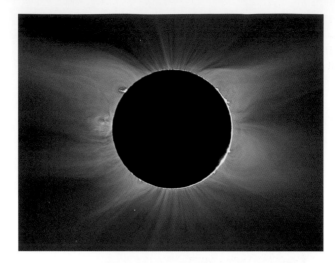

Fig. 11.18. Same details as for Fig. 11.17. In this picture Miloslav Druckmüller has processed Fred Espenak's 2005 eclipse images to maximise the structure of the inner corona in this rare hybrid eclipse where Sun and Moon were almost the same size. © 2005 Miloslav Druckmüller, Fred Espenak.

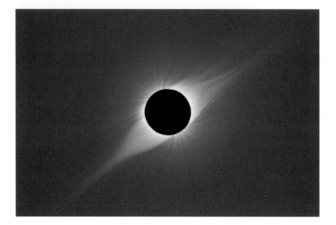

Fig. 11.19. This 1995 eclipse, observed mainly from India, exhibited the classic long East-West streamers typical of a solar minimum corona. The original photographs were taken by Vojtech Rušin from a site at Nim Ka Thana in Rajasthan, India. A 500 mm f/8 lens and Kodak Ektachrome 100 Plus film was used for the eight exposures (1/30 to 4 s). Once again, the image processing on the scanned slides was performed by Miloslav Druckmüller. © 1995 Vojtech Rušin, © 2004 Miloslav Druckmüller.

adaptive mathematical filters which could mimic the appearance of the corona to the human eye. The results have been stunning and Miloslav encourages any amateurs who have good images or photographs of the solar corona at totality to make contact with him. Miloslav's technique typically involves starting with at least eight different coronal exposures, progressively revealing the bright inner and then fine outer detail. Enormous care is taken to rotate and translate (x and y shift) each image to achieve sub-pixel accuracy registration and stacking. Miloslav's own software (*PhaseCorr 2.0*) uses Fourier transform and phase correlation techniques just to get the rotation right. Even when using an equatorial mount this may shift fractionally due to polar misalignment and most amateurs would simply ignore it. The translation registration is achieved by a software package called *Sofo Acc 5.0*

Analyzer and will use a star of, typically, fourth magnitude or brighter, or fine coronal/prominence detail itself, to register the images. Finally, the most powerful technique is to compress the 16 bit data from the stacked images into 8 bits (a reduction from 65,535 to 256 brightness levels) so the computer monitor can display it and the eye view it. The mathematical filters used to simulate the eye's remarkable ability to cope with viewing all the detail in a real life visual solar eclipse are called adaptive filters and the process is sometimes called adaptive convolution. Miloslav's own software, *Corona 2.0*, performs this task beautifully and increases the contrast on fine detail (high spatial frequencies) but without the lunar disc edge "ringing" normally seen with traditional Fourier transform or adaptive convolution software. This attention to detail when registering the images, plus Miloslav's custom software, designed just for coronal processing, produces the most staggering corona visualisations ever made from solar eclipse images.

Older Corona Techniques

Some eclipse photographers, prior to the digital era, captured the corona with colour negative film and then copied the negative onto slide film using filters to remove the orange mask of the colour negative. This technique meant that the large dynamic range of the negative captured the corona well and the projection presentation of a slide conveyed the full beauty to an audience. Thus, the limited ranges of both the slide film and the hard copy printing process were avoided. Another technique practiced by real die-hard eclipse chasers was to actually modify the camera/telescope arrangement such that a metal disc was suspended (using fine wires) above the film plane. Using this method the cones of light going from lens to film are severely vignetted for the inner corona, by the suspended disc, but not vignetted at all for the outer corona. In the UK Dr. Francisco Diego of University College London is perhaps best known for his work in this area. Mike Harlow, of Orwell Astronomical Society, Ipswich, quotes the following formula for making such a suspended disc, on that society's web pages (http://www.ast.cam.ac.uk/~ipswich/Observations/Eclipse-1998-02-26/RD.htm).

The disc diameter and distance in front of the film can be calculated from the following equations:

$$d = D(B + A)/(2D + B - A)$$

$$s = F(B - d)/(B + D) \times 1.8,$$

where d is the diameter of the opaque disc suspended in front of the focal plane at a distance s, D is the diameter of the refractor lens, F is the focal length of the refractor, $A = Ds \times 0.99$, where Ds is the linear diameter of the Sun's image at the focal plane, $B = Ds \times 3.92$.

In Mike's case his telescope had a focal length of 890 mm and the opaque disc was 17.7 mm in diameter and 111 mm in front of the film plane.

However, this technique requires absolute accuracy in aligning the camera/telescope on the corona centre, i.e. the Sun (NB preferably *not* the black lunar disc, which moves during totality): the Sun has to be precisely in the middle of the field. A similar approach adopted by some was to use a radially graded glass filter suspended just above the film plane; again this required much skill in positioning the filter before totality starts, using a precision equatorial mounting.

You may come across the term "Newkirk Filter" in this context. Named after Gordon Newkirk, the former director of the High Altitude Observatory, this is a precision radially graded filter created by evaporating a metal film onto glass in a vacuum chamber, i.e. the same technique that is used to aluminise telescope mirrors. The Newkirk filter attenuates the light from the corona by a Neutral Density value of about 3.1 (1,250×) at the solar limb (which will be hidden by varying degrees during totality) out to a radius of about 10 arcmin from the limb. Its transmission then increases rapidly such that at about 18 arcmin out the attenuation is N.D. 2.0 (100×), at 25 arcmin it is N.D. 1.0 (10×) and at almost 40 arcmin it is N.D. 0.5 (3×). At just over a degree out from the solar limb (i.e. four solar radii) the transmission is 100%, i.e. the filter is clear. Obviously to fully simulate the dynamic range of a Newkirk filter, using non-filtered images, a range of exposures spanning roughly 1,250× would be needed, e.g. from 1/1,000 to 1 s. For more information on exposure values (EVs) at specific ISO values and *f*-ratios, read on! Newkirk is not the only person to have experimented with this sort of technique; as well as Dr. Francisco Diego in the UK, Marius Laffineur and Serge Koutchmy in France have experimented with radial filters. However, with DSLRs, image processing software, and the gigahertz processors now being used, these filters are not essential for good coronal photography. A range of exposures can be taken and mathematically combined to simulate the effects of a radial filter.

Originally, film-based techniques were developed, throughout the 1980s and 1990s, employing a huge range of exposures of the corona; i.e. short exposures for the inner region and long exposures for the outer region. Using darkroom techniques and many, many, hours of hard work, different negatives could be stacked to bring out the maximum detail. The photographic unsharp mask technique, where a blurred out-of-focus positive mask and the negative itself are composited (enhancing fine detail at the expense of coarser over-exposed detail), was used to advantage. However, the techniques began to swing in the digital direction when affordable film scanners appeared in the 1990s and when home computers became fast enough to manipulate millions of pixels of data in minutes rather than days. Eclipse negatives and slides could then be scanned, different coronal exposures stacked, and the output printed by specialist photographic printers who could accept digital files. Looking back through my copies of the excellent astronomy magazine *Sky & Telescope* it would appear that messrs Albers, Shiota and Ressmeyer were the pioneers in this field around 1994/1995. Since the year 2000 digital SLRs have revolutionised the entire process by providing multi-megapixel data files which can be transferred in seconds to a PC and processed in *Adobe Photoshop*, *Paintshop Pro* or any other suitable package. Suddenly, bringing out detail in the corona has climbed new heights and every amateur can produce stunning solar corona images with care. However, the basic technique is the same at the eclipse site. You take loads of exposures at different shutter speeds so you have a set of images optimised for every part of the corona as well as the shortest one for the prominences. After the eclipse is over you can have great fun (and frustration) in the coming hours, days and weeks producing your masterpiece! If you want to reproduce an image as near as possible to the view you saw through binoculars, make sure you do a lot of visual savouring during totality. Indeed, you may wish to make an audio recording describing what you saw, or simply describe the visual view on your camcorder's sound track, especially noting the colours, which can easily be distorted when digitally manipulating images.

Practical Coronal Processing

So once you have a set of digital images at various exposures, revealing details from the inner corona (shortest exposures) to the outer corona (longest exposures) how do you combine them? In practice your precise methods and keystrokes will depend on what image processing software you own. Many lunar and planetary imagers simply own *Adobe Photoshop*, *Photoshop Elements* or *Paintshop Pro*, along with Cor Berrevoet's *Registax* stacking software. More advanced imagers use custom astronomical software such as *Maxim DL* or Richard Berry and James Burnell's *AIP* software. So it is impossible to give a precise key-press by key-press tutorial that will satisfy everyone. However, whatever software you use, the basic principle is to brighten the faint outer corona and suppress the brilliant inner corona while preserving fine detail in both. Half of this battle can be achieved simply by stacking a set of images at different exposures and averaging the end result. For example, imagine a set of eight coronal exposures with a DSLR set to ISO 400 and a 400 mm f/8 lens. The exposures chosen are, for example: 1/4, 1/8, 1/15, 1/30, 1/60, 1/125, 1/250, and 1/500. The inner corona will be a total whiteout in the 1/4 s exposure but the maximum value cannot exceed the 8 bit (255) maximum value per colour channel so the overexposure is limited by saturation. Simply adding and averaging all these eight images in, say, *Registax*, will capture a far greater dynamic range from the corona, as an exposure range of 125× has been compressed into one image, with the saturated inner corona contribution from the deepest image being reduced by a factor of eight. However, we can be far more adventurous than this. The real trick is to enhance fine detail and suppress large brightness variations within each image prior to stacking and averaging, thus compressing the huge coronal brightness variations for each exposure and for the stacked result. The most powerful single technique to use here is the unsharp mask filter which can be found in virtually all astronomy and image processing software. In *Photoshop* it appears under the Filter/Sharpen menu and in *Paintshop Pro* it lives under the Adjust/Sharpness menu. The wavelet sliders in Cor Berrevoet's *Registax* perform a similar function. Playing about with all the settings for this filter should result in a dramatic enhancement of coronal detail in each image, from the shortest to the longest. It is important with any astronomical image processing to tweak and play with everything until you are satisfied. Using a log power law filter can also suppress the bright detail and enhance the faint detail too. Using such a filter before the unsharp mask is worth a try! Once you feel you have tweaked every individual image to perfection you can then stack them up in, for example, a freeware package like *Registax* or *Iris*.

Digital photography using high dynamic range (HDR) techniques is gaining popularity in landscape photography with so many people now using digital cameras and powerful image processing software. In daylight scenes where objects do not move between exposures a number of images can be taken to mimic the abilities of the human eye. In other words, short exposures can capture bright clouds, while longer exposures capture details in the shadows. *Photoshop CS2* has an HDR feature for this very purpose and there are also some custom packages such as PhotomatrixPro by *HDR Software* (http://www.hdrsoft.com) and FDRTools (Full Dynamic Range software from http://www.fdrtools.com). If a pre-scripted set of exposures can be sent to the camera during totality using camera control software like *ImagesPlus* more time can be spent observing visually while a whole range of corona exposures are captured.

Radial Blur Subtraction

One corona processing approach that has become quite popular in recent years is sometimes called the "Pellett method" as it was first brought to the attention of eclipse photographers by Gerald L. Pellett in the January 1998 edition of *Sky & Telescope*. It is worth bearing in mind that 300 MHz processors were the state of the art in 1998 and astronomical image processing software was in its infancy. Nevertheless Pellett recognised that if you used *Adobe Photoshop's* radial blur filter (Fig. 11.20) on an image with the Sun at its centre, you could create a useful unsharp "mask" of the low frequency brightness variations in the solar corona. Think, for a moment, about what we are seeing in the corona. It is a largely fuzzy whitish mush with some subtle, but exquisite, fine detail superimposed on the mush! The eye can pick out the subtleties well, but in a straight exposure it looks pretty featureless. Also bear in mind that the fine details are, in general, radiating outwards on extended radii from the solar centre. Very few fine details in the corona are tangential to the solar disc. *Photoshop's* radial blur filter, typically set to 10° in Pellett's example, creates a seamless series of smoothly overlapping sectors around the Sun which have blurred the coronal brightness in each radial line to the average brightness in each 10° sector. Thus, all fine detail is blurred out leaving just the general low resolution radial coronal variations. Just what we DO NOT want you might say! Yes indeed, but this blurred mask is then *subtracted* from the original so only the really high frequency, fine detail remains. Clever eh? This is

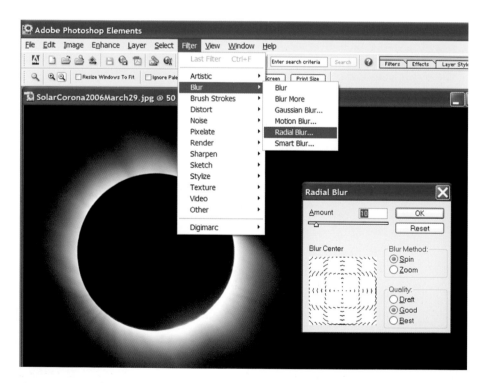

Fig. 11.20. As described in the text, the radial blur tool of *Adobe Photoshop/Photoshop Elements* is a powerful tool for enhancing radial coronal detail.

actually the basis of all unsharp mask processing but the critical factor here, that Pellett exploited, was the fact that *Photoshop's* radial blur filter was ideal for the job, because coronal detail is largely radial. The precise *Photoshop Elements* (a less expensive version) key presses required to get to this stage are as follows (assuming the Sun has been placed in the centre of each image):

Load corona image (File-Open); filter-blur-radial blur (Spin method and Good quality setting) to create the blurred mask (set to 10°); save the blurred image (File-Save as). Then, call up the original image so both images are visible. With the radially blurred image as the active one "Select-All" and "Edit-Copy" and then make the original image the active one. "Edit-Paste" to transfer the blurred image into the original and then click "Window-Layers" with the default blending method changed from "Normal" to "Difference", i.e. subtract the blurred image from the original. Playing with the opacity value here gives you a variety of appearances for the "Difference" image. Selecting "Merge Down" merges the subtracted image layers into a final image, which should be saved at this point. The resulting image may well be rather weird and black but it should reveal all the intricate fine radial detail in the image. Pellett recommended using an offset of 128 (scale of 1) to brighten the very dark image. By playing with *Photoshop's* brightness and contrast tools you can make the difference image into something less black. What I have painstakingly described above is the production of a *single* radially enhanced corona image using *Photoshop's* powerful radial blur tool. In practice the eclipse-chasing image tweaker will use many radially blurred and subtracted "difference" images from various exposures and stack them up (e.g. using *Registax* from http://registax.astronomy.net/). Plus much brightness/contrast/ colour tweaking will doubtless be carried out too. In the original Pellett *Sky & Telescope* article the multiplication of the pixel brightness values in the stack of subtracted images, by the pixel brightness values in the longest exposure was recommended for a greatly detailed result. In the full *Photoshop* package most of this is carried out using the "Image-Apply Image" command. Alternatively, multiplying each subtracted image's pixel values by the pixel values in the corresponding original, and then stacking those images was recommended for a more "visually realistic" approach. However, in 2008, with far more astro-software and computing power available, not least Cor Berrevoet's *Registax* and its wavelet sliders, I suspect many amateurs will enjoy dabbling with that software and other tools. Nevertheless *Photoshop's* radial blur filter is a very powerful weapon for extracting coronal detail, although some radial enhancing comet "filters" will produce a similar outcome if you have the specialist software.

UK amateur astronomer Pete Lawrence provided me with the steps he used for enhancing his own coronal shots from the 2006 Total Solar Eclipse. The critical stages in his own version of the Pellett technique are shown and described in Figs. 11.21 and 11.22. A similar approach, yet again using *Photoshop's* radial blur feature, was described in the May 2006 issue of *Sky & Telescope*, by Maurice Hamilton. In Hamilton's method Photoshop's rulers and guides were used to form a grid in which the lunar silhouette could be aligned. (NB bear in mind though that the Moon moves with respect to the corona; see below).

Using *Registax* on its own to wavelet enhance each coronal image, from shortest to longest exposures, and then to also stack the resulting images and average them will produce a dramatic result not much different to the radial blur method. In practice, whichever technique you use, endless tweaking and experimenting will really pay dividends. Every total solar eclipse is different as is everybody's

Fig. 11.21. This sequence of three images shows the application of the Pellett technique (see main text), i.e. subtracting an eclipse image subjected to Photoshop's Radial Blur technique from the original and then using the difference image as a multiplier with the original. The images were exposed by Pete Lawrence of Selsey, UK who was in southern Turkey for the 29 March 2006 total solar eclipse. He used an 80 mm aperture f/7.5 Skywatcher 80ED Pro refractor with a 0.63× telecompressor (giving f/5, i.e. 400 mm focal length). A Canon 10D with 7.4 micron pixels was used, i.e. an image scale of 3.8 arcsec/pixel. The original mid/outer corona shot at top was shot with a 1/60 s exposure at ISO 400. Below that is the radial blur difference image (slightly enhanced to reproduce

(Continued)

equipment and set of images. As I have already mentioned, *Photoshop* CS2 has a useful HDR feature which is of great interest to all corona photographers.

For absolute perfection in coronal photography it is worth bearing in mind the drift of the Sun, Moon and corona across the sky during the exposure (i.e. if you are using a fixed tripod) and also the drift of the black lunar silhouette, over several minutes, relative to the corona (even if you are using an equatorial head). The Sun moves at roughly 15 arcsec/s of time across the sky. With typical portable equipment and an aperture of 100 mm you will only resolve arcsecond detail in the corona. However, that does mean freezing any drift to about half an arcsecond during the exposure. On a fixed tripod that means exposures no longer than 1/30 s, if we are aiming at perfection. In practice though, the perfectionist will use a portable lightweight equatorial head, like Takahashi's Sky Patrol/Teegul mount. The other issue is a bit more subtle to grasp. Even if the fine coronal detail is frozen in each image, if you take coronal exposures over, say, several minutes, i.e. in a reasonably long totality, the Moon's black disc will drift with respect to the coronal detail. The drift is quite significant, roughly half an arcsecond per second of time. This might not seem important if the individual exposures are short. However, when you are stacking multiple frames on top of each other you have to stack them relative to something. If you use the black lunar disc as the reference the coronal detail will be badly smeared in the stacked image, even with a time span of 10 or 20 s. If you use a feature in the coronal detail as a reference the lunar disc will drift into an ellipse in the stack. Most experts in this field will manually align the coronal detail and then put an artificial black disc over the multiple stacked Moon discs. Cheating? Well, maybe! A simpler alternative, if you are on the centre line and using a modest focal length lens, would simply be to take all your coronal pictures within a few seconds of mid-totality (maybe with a wide auto-bracketing setting) and use the lunar silhouette as the reference. Some amateurs take the ghostly lunar Earthshine detail (the lunar surface illuminated by sunlight reflected from the Earth) from their very longest coronal exposure and paste that image where the Moon is, for a really dramatic final composite. This may require an exposure of many seconds, depending on your *f*-ratio.

Exposure Times

As we have seen, quoting precise single exposure times for solar eclipses is fraught with danger, and bracketing exposures for every type of photograph is the only sensible approach. In the tables at the end of this section I have attempted to

---

better in this illustration) and, at bottom, the final enhanced result created by overlaying/blending the difference image data by the original. Note, this is just applied to a single image. The precise steps, supplied by Pete Lawrence were:

To each image . . .

– Radial blur of 10 with quality set to best was applied to a duplicate of the original image.
– Original was then selected, duplicate layer created and Photoshop Apply To this duplicate layer. Radially blurred image was subtracted from the duplicate layer – opacity 100, scale 1, offset 128.
– Original layer was then duplicated again and moved on top of previous result. A blend mode of overlay then brings out the detail.
– Repeat for ISO 400: 1/750 s, 1/180 s, 1/60 s, 1/20 s + ISO 800: 1/30 s (see Fig. 11.22).

Image: Pete Lawrence

Fig. 11.22. Observer and equipment details are identical to Fig. 11.21. Except in this case (*top*) we see the result of using the Pellett technique on five images from inner to outer corona (1/750 to 1/20 s), stacking them and then applying a further radial blur subtraction/multiplication on the final stack. The precise steps, supplied by Pete Lawrence, following his steps in the previous figure) were: Layer each frame with the shortest exposure at the bottom of the pile.

Hide all images except the lowest one. By showing each image above the lower one in turn and temporarily making it semi-transparent, make sure all images are aligned with one another.

Working on a pair of images at a time from the bottom up . . .

– Hide the upper layer.
– Select the lower layer and draw a circular selection around the part of the lower layer you want to keep.
– Copy this to the clipboard with Ctrl-C.
– Show the upper layer, make sure the layer is selected and press Alt then click on the layer mask button at the bottom of the layer floating menu. This pastes the clipboard contents into a layer mask for the upper image.
– At first the edges of the mask will appear defined. Apply a Gaussian blur to the mask until the edges disappear and the blend from one layer to the next is seamless.

(Continued)

suggest starting points for various eclipse targets based on an ISO 400 camera setting. However, these are only starting points and poor transparency, thin cloud and different types of equipment can drastically alter the optimum exposures. I am, of course, recommending that the camera is set to manual exposure mode where you choose the exposure and you choose the best point of focus too. Cameras are designed to cope with everyday events and solar eclipses are highly unusual. It is unlikely, though not impossible, that a camera's auto-exposure mode will get the exposure right where totality is concerned and, of course, it will not know that you want to take a range of exposures from the diamond ring/Baily's Beads ultrashort ones through to the long outer corona shots. Of course, if you are going to attempt a complete HDR corona masterpiece you will, as just described want to take loads of different exposures. On the day you need to already be instinctively familiar with your camera's manual exposure mode and (if using a lens) manual focus mode too.

Recommended Exposure Times for Specific Lens *f*-Ratios

ISO 400 Settings Throughout, to Minimise Exposures: For ISO 200, Double the Exposures, and for ISO 100, Quadruple the Exposures

In practice a range of exposures slightly shorter and slightly longer should be attempted.

Where a digicam is pointed down the eyepiece of a telescope of a specific *f*-ratio these figures may well be *very* wide of the mark and exposures will depend on the eyepiece and degree of zoom used. But they may still be useful as a rough starting point. If in doubt use every shutter speed you have! For sharp pictures with cameras mounted on lightweight "travel friendly" tripods, exposures of 1/500 and faster will be necessary to freeze fine prominence details. For Baily's Beads and the Diamond Ring much slower ISO settings may be best, especially if your camera does not allow exposures shorter than 1/1,000 s.

Wide-field panoramic twilight shots during totality

f/1.8	f/2.8	f/4	f/5.6
1/16	1/8	1/4	1/2

Solar disc partial phase with a solar filter (ND 5)

f/8	f/11	f/16	f/22
1/500	1/250	1/125	1/60

Close-up outer corona

f/8	f/11	f/16	f/22
1/8	1/4	1/2	1 s

High resolution chromosphere/prominences

f/8	f/11	f/16	f/22
1/2,000	1/1,000	1/500	1/250

High resolution Baily's Beads/Diamond Ring

f/8	f/11	f/16	f/22
1/16,000	1/8,000	1/4,000	1/2,000

Running another Pellett process over the final composite (flattened of course) can also bring out some subtle detail.
In the final bottom image a 1 s exposure revealing Earthshine detail has been pasted onto the dark lunar silhouette. Image: Pete Lawrence

Multi-Sun Exposures

Many eclipse chasers like to try a "multi-Sun" set of images (Fig. 11.23) showing the Sun being gradually eclipsed, with a shot of totality in the middle and then the partial phases at the end. Successfully completing a multi-Sun exposure set assumes that you will arrive at the eclipse site well before first contact and depart after fourth contact. The former may be impossible if the roads near the eclipse site are jammed and the latter may also be thwarted if the majority of people on your tour want to leave before fourth contact. Let's face it, the period after totality is a bit of an anti-climax!

Prior to the era of digital film scanners and digital cameras a multi-Sun image involved meticulous planning and multiple exposures on one piece of film. It can be horrendously difficult to plan such an exposure even if you practice the feat at home before you go. Plus, one mistake on the day would trash the attempt. The curving path the Sun takes across the sky in the hour before totality, and the hour after totality, can totally confound you when your latitude is unfamiliar. It is not uncommon for eclipse photographers to suddenly find the Sun is heading out of the frame far earlier than expected. The Sun can travel in a 40° curving arc from first to fourth contact necessitating a fairly wide angle camera lens. This, in turn, will mean that the individual Sun images are tiny and show little detail. So you may prefer to capture a smaller range of eclipse phases at a higher focal length. So far I have been describing the old-fashioned fixed camera approach where you guard your camera and tripod with your life in case someone walks into it or through the field of view. The most ambitious multi-Sun pictures of the film era often contained a spectacular building or monument in the foreground to enhance the drama; but they were nerve-racking to set up! However, in the digital era, or even with a film scanner, things are far less fraught. You can take loads of different exposures of each phase of the eclipse and simply combine them digitally at a later date. If the camera is knocked, or cloud intervenes there is no problem; you can sort it all out at a later date. The same applies for the spectacular foreground object. All you have to do is get a good picture of the building, church, pyramid, waterfall, etc., and clone brush it into the final picture.

Fig. 11.23. A multiple exposure Sun image (multi-Sun) from the 11 August 1999 Total Solar Eclipse by Nigel Evans, at Sivas, Turkey, using a 250 mm lens. Before the digital era these sort of shots took considerable advanced planning. Image: Nigel Evans

Capturing the Shadow
with a Fish-Eye Lens

Another challenge often undertaken by eclipse chasers is to grab a record of the entire sky as the shadow of the Moon sweeps over the eclipse site (Fig. 11.24). For such photography a Fish-Eye lens is often used, although these lenses do not always cover the entire sky, especially if you are using a two-thirds format digital sensor. Traditionally, with a 35 mm format SLR film camera (or a full frame DSLR) the 8 mm focal length lenses put the entire hemisphere of the sky, i.e. 180°, onto a 24 mm wide circle on the film (or on the detector). The ends of the 36 mm long frame are not used and are just black in the resulting picture. With the 16 mm focal length lenses, which are also called Fish-Eye lenses, the 35 mm format film/sensor diagonal covers 180° so quite a bit of the sky is lost. Obviously with the two-thirds (23 mm × 15 mm) CCD or CMOS sensors found in the lowest priced DSLRs (such as Canon's Digital Rebel/300D/10D) even less field will be captured.

One good source of really inexpensive Fish-Eye lenses is, yet again, the Russian Zenit company which has dealers in most countries. For $250 their 8 mm f/3.5 Peleng Fish-Eye can be purchased brand new, at a fraction of the cost of a Canon or Nikon lens. However, some caution is needed. Firstly, make sure that the lens will fit your camera body. Many Russian lenses, new or used, have a 42 mm Pentax thread which will not be compatible with a modern bayonet mounting. Also, the rear lens of most Fish-Eye lenses intrudes into the camera throat by quite a distance. With some cameras there is simply not enough room for the reflex mirror to move upwards past the Fish-Eye's rear lens. The lens manufacturer Sigma makes 8 mm Fish-Eye lenses for around $600 and they are compatible with modern

Fig. 11.24. An excellent 16 mm fish eye image of the umbral tunnel at the very low altitude Australian Outback Eclipse of 4 December 2002. Image: Nigel Evans

Canon and Nikon bayonet mounts. A number of companies market so-called "Fish-Eye converters" but these add-on lenses effectively shorten the focal length of the existing lens by, at best 0.4×. So, to simulate an 8 mm Fish-Eye you would need to be using a very wide angle lens, of around 20 mm focal length, to start with; plus, you would need to be using a full frame 36 × 24 mm sensor or good old fashioned film. Enterprising amateur astronomers and eclipse chasers have devised (or, rather, bodged!) other Fish-Eye solutions together though. Those bowl shaped blind-spot mirrors that you see in multi-storey car parks and concealed entrances have been hauled into service. A separate camera on a tripod above the mirror, which is placed on the ground, is used to photograph the all-sky image (plus the reflection/silhouette of the camera and tripod). People have even attached those door peephole viewers in front of digicam lenses to record a fish-eye scene, but, rather obviously, the image quality is considerably inferior to a proprietary camera lens.

The Fish-Eye approach does offer a unique way of photographing the lunar shadow passing overhead though. If a series of images are taken at, say, 15 s intervals, a dramatic animation of the eclipse sky can be captured. Obviously the right choice of exposures is essential and, yet again, exposure bracketing is essential. However the values I have indicated in the table do provide a good starting point.

Flash Spectrum Photography

During a total solar eclipse, an exceptional opportunity to observe an emission spectrum occurs (see Fig. 11.25). The opportunity does not last very long though and it is right in the most panic-stricken periods, at second and third contacts. The normal solar spectrum, observed without the need for a total eclipse, is the result of the light from the Sun passing through the relatively cool 5,700 K solar atmosphere. This atmosphere leaves absorption lines in the spectra and these lines give the star's characteristic signature. This principle applies whether the star is our

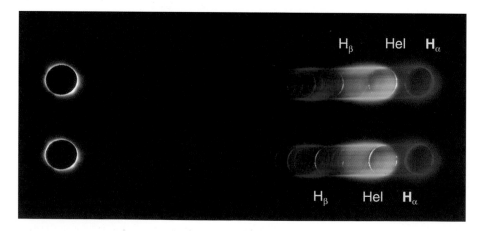

Fig. 11.25. The Flash Spectrum at both second and third contacts at the Australian Outback Eclipse of 4 December 2002. A 250 mm lens plus diffraction grating. Image: Nigel Evans

Sun or another star, or a nova or supernova being studied with a large telescope. When the last chink of the bright solar disc or photosphere is covered by the Moon at second contact the relatively thin pink sliver of the chromosphere is visible, just for a matter of seconds, before it too is covered by the Moon. The reverse sequence occurs at third contact. At this brief period the light from the brilliant photosphere is hidden and the light from the much hotter (10,000 K) chromosphere shines on its own without being swamped by the brilliant photosphere. The chromosphere is a hot gas of low density and so emits an emission spectrum featuring bright lines at wavelengths corresponding to the atoms actually making up the emitting gas. This spectrum can only be recorded at second and third contacts. Another bonus is that to obtain a spectrum of anything a narrow intense slit of light is desired. Normally spectroscopes are designed with their own man-made slit, but in this case the thin skin of the chromosphere, seen edge on and shortened into a brief arc by the Moon, provides a natural slit. So the only device needed to produce a spectrum that a camera can capture is an inexpensive diffraction grating, ruled with a few hundred lines per millimetre. These are obtainable from suppliers such as *Rainbow Optics* or *Edmund Scientific*. Typically, the grating is placed in front of a camera lens of 200 or 300 mm focal length and the eclipsed Sun is placed in one corner of the field of view. The grating and camera are aligned such that the flash spectrum will appear as a line, some distance from the image of the eclipsed Sun, stretching across the picture. The blue part of the spectrum will be nearest to the direct eclipse image and the red part will be the furthest, but still (hopefully!) on the frame. Obviously considerable experimentation is required in the weeks and months before the eclipse to work out how to orient the grating and where to place the Sun in the frame. Some observers have employed an offset finder telescope on the camera arrangement so they can centre the Sun precisely. A smaller focal length lens will make centering the whole Sun and spectrum easier but at the expense of resolution in the spectrum. Potential test objects for the experimental set up can be *indirect* Sunlight glinting off a very distant shiny object (as always, take great care whenever the Sun is involved) or a mercury or sodium streetlight at night. Bear in mind that the sort of exposures you will need for recording the flash spectrum will typically be dozens of times longer than for recording the chromosphere and prominences without a grating with the same equipment. The best way to obtain the correct exposure is, again, to take numerous exposures and to duplicate someone else's equipment, techniques and exposures that were successful on a previous eclipse trip.

If your Flash Spectrum experiment works you will end up with a spectrum of the chromospheric arc, each colour curved, like the arc itself, on the film. If you are at an eclipse where the Moon is only just big enough to cover the Sun, the spectrum will consist of myriads of overlapping coloured solar rings, as the chromosphere will extend all around the lunar disc, for a few seconds. You will notice that, every so often in the spectrum, there is a pronounced coloured arc, far brighter than surrounding arcs. These are the arcs produced by specific bright emissions in the chromosphere and are easy to identify. The arcs may also look irregular, due to lunar mountains near the limb and you may find horizontal bright streaks crossing the spectrum in some exposures. These streaks are simply pollution from the dazzling photosphere as some of its light may still seep through lunar valleys. The brightest arcs in your spectrum are from the red Hydrogen Alpha line at 656 nm, the yellow Helium line at 588 nm and the blue Hydrogen Beta line at 486 nm. More subtle peaks might also be detected by careful analysis, notably the green 530 nm iron line and, if you have a super quality spectra, a deep red Helium line at 707 nm, a blue Hydrogen Gamma line at 434 nm and violet Calcium H and K lines at 393 and 397 nm.

Automating the Photography

I have already stressed, many times, the frantic nature of solar eclipse photography. However, until you have experienced those fleeting moments of totality, and time passing five times quicker than normal, you will not really appreciate the experience. Indeed, even after you have seen a total solar eclipse and wonder where the time went, the brevity of the experience tends to fade once you return home and the day job and domestic chores take over your mundane, sad and tragic life. However, a few determined individuals, after each eclipse, promise to "be prepared" the next time around and with an iron resolve they decide to automate and customise their equipment for the next eclipse trip. To my knowledge no-one markets dedicated eclipse photography equipment so here is an opportunity for the individual to make something unique so that the next eclipse will be a far better ergonomic experience with more time for savouring the view. The good old amateur "build it yourself and be proud of it" tradition can still survive even in this "off the shelf" telescope era.

The ultimate eclipse photography system would allow the observer to lie back in a deckchair with a pair of binoculars (or a small telescope) while a collection of cameras simply fired away on equatorial mountings, with no human intervention. The only role of the observer would be to whip all the solar filters off 30 s before second contact and put them all back on shortly after third contact. For the dedicated eclipse photographer with, typically, a couple of long focal length camera systems (one for prominences and one for whole corona shots), a flash spectrum camera, a multi-Sun camera and a fish-eye panorama camera, having five (or more!) cameras under remote control is utter bliss. However, as veteran eclipse chasers will only be too well aware, automated systems do not always work flawlessly. Unless the system is 100% tried and tested the amount of hassle can be multiplied, not reduced. Even removing five solar filters can be a fraught affair, especially if you stumble over a camera in the fading light. Of course, if you have your cameras controlled centrally, from one palmtop PC, or a laptop, there will also be a myriad of wires waiting to trip you up.

It is particularly important that you know the precise start and stop times of totality from your observing site and that you have an easily visible clock which is accurately set to U.T./G.M.T. On many of the eclipse trips I have been on, we have been mobile, in a convoy of coaches, up to an hour before totality. Sometimes things are cut this fine because organisers are searching for a cloud gap. At other times it is because roads are even more congested than expected, with gridlock within 10 km of the proposed site. If you do not end up precisely on the centreline the period of totality may be much shorter than you expected. I will quote the formula again, to save the reader some effort! The formula is: Duration = Centreline Duration $\times \sqrt{[1-(2D/W)^2]}$ where W is the width of the track and D is the distance from the centreline (at right angles to the track). Second contact may be much later than expected and third contact may be much earlier. This is bad news if you have inflexible software controlling your cameras that is set for a precise set of exposures, e.g. short diamond ring/Baily's Bead/prominence exposures at second and third contacts and long corona exposures throughout totality. Some intelligent planning is required so your software can cope with any eventuality. For example, you may want the ability to input new second and third contact times into the program, or, devise a powerful piece of software that can just take your new latitude and longitude into account. You may well want to consider taking a GPS unit on holiday so you can have no doubt about where you are on the Earth's surface during totality. As a minimum, if your software is not flexible, you should allow a wide timing window in which to shoot the crucial, short, second and third contact images, so that even if your timings are 10 or 20 s adrift you will still be taking short exposures at those crucial times. I am, of course, assuming the totality period is of a much longer duration than those windows!

Commercial/Freeware Solutions

Before you hold up your hands in horror at the thought of writing software, and building hardware, to automate your eclipse photography let us first consider what can simply be purchased or acquired to make life easier. An absolute must for any non-automated astro-photographer is a remote shutter release to avoid any vibration from the observer's trigger finger. Canon's RS60-E3, already shown in Chap. 9 (Fig. 9.5) is straightforward and inexpensive. Fortunately, in this DSLR era some quite sophisticated programmable remote control options are available too, even if there is nothing that is perfect for eclipse photography (at least, not at the time of writing, in 2007). The Canon TC80-N3 remote controller comes part of the way to easing the eclipse photography problem. As well as featuring a self-timer delay of up to 100 h (in 1 s increments) it has an interval timer (intervalometer) which can arrange for pictures to be taken with a fixed interval between them. So, for a 200 s duration total solar eclipse you could, say, order it to take 100 shots every 2 s, if your camera and memory card can sustain that rate. Sometimes it can be tricky to get exactly what you require when combining a remote controller with a camera. For example, most DSLR's feature an auto-bracketing mode whereby if you take a picture the camera will automatically expose at the desired/optimum shutter speed as well as at a half, a quarter, double or quadruple that exposure. (In photographic jargon half or double represents one photographic stop or one EV). Finer bracketing divisions are selectable with many cameras. The tricky bit can be combining auto-bracketing with the use of a remote controller like the TC80-N3. You may well find that you need to set the camera itself to self-timer and auto-bracketing and then set the interval timer (and exposure counter) on the TC80-N3. As I've repeatedly stressed in this book, you need to practice all these techniques weeks before the eclipse and certainly not on eclipse day!

As well as programmable remote controllers for each camera there are a number of items of software which can be usefully employed for controlling DSLRs. Many cameras come with basic remote control packages which can talk, via a PC, to the camera down the USB/*Firewire* link that interfaces to the camera for basic image downloading. As a minimum such software can be used for basic remote operation of the camera, with the big advantage that images can be inspected at high resolution on a lap-top screen to make doubly sure that the camera is precisely focused during the partial phases just prior to totality.

Mike Unsold's *ImagesPlus* software package is proving popular with astro-imagers who use Canon and Nikon DSLRs. This is a comprehensive package which is aimed at amateur astronomers who want to convert, edit, smooth, align stack and image process large DSLR images and who want to control their DSLR from their PC. However, of most interest to eclipse chasers is that it can control Canon and Nikon DSLRs from a pre-determined script of up to 30 exposures with user-defined exposure times and ISO settings. Many amateurs used this package successfully at the 29 March 2006 total solar eclipse. With any automated software of this nature it is important not to overload the camera or software memory buffer by asking it to take too many large images in very quick succession. Thus a practice run is essential before the big day. It may be necessary to put 1 or 2 s delays between exposures, in the script, to prevent this occurring. The *ImagesPlus* software website is at http://www.mlunsold.com/.

Breeze Systems have produced a software package called *DSLR Remote Pro* that may be of interest to some eclipse chasers. This package allows full control of a DSLR from your laptop, with the emphasis on auto-bracketing of up to 15 shots, for coping with a huge brightness range; something that could be very useful for corona photography. It also has a time-lapse feature which allows the number of photos in a sequence and the interval between shots to be specified. More information is available at http://www.breezesys.com/DSLRRemotePro/index.htm. Bearing in mind the number of focusing problems people experience at total solar eclipses the software *DSLR Focus* by Chris Venter (http://www.dslrfocus.com/) may well be of interest, especially if, you are using a small portable telescope with an ASCOM language compliant focuser which you plan taking to the eclipse. *DSLR Focus* is aimed at amateur astronomers, not just eclipse chasers, and enabling auto-focussing of DSLR's on starry targets is only one of its functions. At the time of writing it has been expanded to control more than just Canon cameras but it specifically works in conjunction with the Canon TC-80N3 to allow capture and download of images from your Canon camera to your laptop/PC.

Custom Solutions

The eclipse chaser Fred Bruenjes (http://www.Moonglow.net/eclipse/photo_software.htm) has also developed some very interesting software too and it can be downloaded from his site. This particular package, called, not surprisingly, "Eclipse" allows the control of up to four, yes *four*, cameras including one Canon DSLR camera which gives you a powerful advantage on eclipse day. Cameras can, assuming all goes smoothly, be dedicated and automated for, e.g. high resolution, flash spectrum, fish-eye and multi-Sun pictures and controlled just from the one laptop. I am not aware of any commercial package that allows you to control more than one camera in this way. Fred's software uses USB/IEEE 1394 (Firewire) control of the DSLR, with full control of shutter, aperture, ISO, and file type/quality for the main camera. For an additional three cameras, e.g. older film cameras, serial port control of the physical shutter press operation is available. In all cases, but specifically for the main DSLR camera, any digital images must be stored in the camera and not on the PC, so *DSLR Focus* or *DSLR Remote Pro* may be of more interest if you place multiple camera control as a lesser priority than focusing and seeing the images on a PC screen at full resolution. Bruenjes software has some nice touches though which only those who have been at a total eclipse site, and experienced the photographer's panic will appreciate. For example, countdown timers are available and you can record an audio reminder to say "Filters Off" to avoid a disaster in the heated moments before totality. It is definitely a piece of software designed by someone who knows what it is like to experience totality when you have numerous cameras to control. I could say more but simply downloading this software from Fred's site is the best way to learn more. For Macintosh users Glenn Schneider has written a useful piece of software called *Umbraphile* which may well be of interest. See http://nicmosis.as.arizona.edu:8000/ECLIPSE_WEB/UMBRAPHILE/UMBRAPHILE.html.

Essentially both these pieces of software, by Bruenjes and Schneider, attempt to tackle the two methods of controlling cameras via a PC or a Mac, i.e. by a simple on/off relay action or by intelligent Firewire/USB control. In the first instance, popular amongst electronic hobbyists in the film era, the most basic operations, such as operating the shutter (and, potentially, crudely adjusting the exposure too) can be controlled on old cameras and even DSLR's simply by using a PC's serial port to control relays that physically, or electronically create a simple on–off action. The precise nature of this action will depend on your set-up and how much you want to modify an old camera. In the simplest case a relay can be made to operate a cable release or close a circuit such that the camera takes a picture. With, say, an old film camera like a Canon T-70 the camera will then advance the film automatically. With a modern DSLR the image will be stored in its buffer and the camera will await a further command from what it thinks is the remote control port, but is actually a line from your PC and a homemade box of tricks. It helps greatly to have some basic electro-mechanical knowledge when making such systems but there are always plenty of people in the astronomy community who are able to give advice with such projects. In recent years serial (and parallel) ports on PCs have been dying out in favour of the much faster USB, high speed USB and Firewire ports. This is a backward step for the gadget builder as the old serial ports were incredibly useful for little projects like automated camera control. However, do not despair, because you can now purchase serial port replicators that fit into a USB port and impersonate a real serial port. In addition, higher specification laptops often come with three or four USB ports and so the possibility of controlling multiple cameras still exists. However, at the time of writing I am unaware of any software that can control multiple modern DSLRs and enable all the camera images to be transferred to the PC. Ideally, this is what is required.

My good friend Nigel Evans, of Ipswich, England, was one of the first observers to build a system for automating a set of eclipse film cameras. In his system an early palmtop type computer, called a Psion 3c (Fig. 11.26), and its optional parallel printer port interface were employed. The eight lines from the parallel port interface were used so that they could control, in theory, up to eight camera shutters or other on–off operations. An electronic component called a 74LS373 Octal D-Type Flip-Flop was used to take the Psion 3c outputs and provide them with far more current capability so they could easily operate some relays which Nigel attached to various camera shutters and function keys. This system was especially handy as the Psion 3c was a very lightweight affair (a godsend for eclipse trips) and its OPL programming language was easy to use.

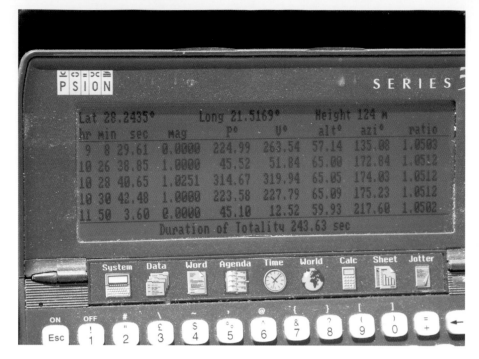

Fig. 11.26. An old palmtop computer (a Psion is shown here) can be coaxed into service for controlling relays to press camera shutters automatically at an eclipse site. This also enables some flexibility if the location changes slightly and if there is software provision for coping with the different timings at a new site. Image of Nigel Evans' Psion taken by the author

Simply Making Life Easier

As well as devising automated eclipse controllers for the really dedicated eclipse nut it is worth investigating a few, less technical, ergonomic solutions to make eclipse day go more smoothly. Indeed, one of the best antidotes to the post-totality anti-climax period is to stroll along the desert, or whatever site you have observed from, just to amaze yourself at the variety and ingenuity of the contraptions on display. People bring all manner of home-made and commercial toys to eclipses, from complete 35 cm aperture telescopes to domestic colanders for creating a multiple pin-hole camera effect. You will even see shadow band detection equipment (camcorders pointed down at calibrated white cardboard discs) and the odd deckchair for those who like to soak up the visual view and do nothing else. Between these extremes there are a few real gems that have been assembled in the garage. Take a look at Chris Baddiley's amazing home made camera mounting in Fig. 11.27 to see what I mean! Some of the very best eclipses tend to occur when the Sun is virtually at the local zenith, which can be a real pain (literally) to observe. Even if you have a Schmidt-Cassegrain, or a Maksutov with a star diagonal viewer, you may well find that the eyepiece is still at an inconvenient position and, in the worst case, that your tripod is tipping over. Binocular users may find it impossible to get comfortable while holding a heavy pair of 15 × 80 s up at the

Fig. 11.27. The magnificent, home-constructed, quad camera equatorial mounting of Dr Chris Baddiley on site at Jalu in the Libyan Desert, awaiting the 29 March 2006 Total Solar Eclipse. The mount has different Declination heads for different projects. As the 2006 total eclipse was only at a Declination of just over 3° the Dec tilt is very small in this case. The polar axis is made from a modified military theodolite axis. A polar axis alignment telescope fits onto the mount. The home made Right Ascension drive can be adjusted to run at any rate and has preset sidereal, solar and slow rates. The Declination tilt adjuster has a motor drive too. The systems mounted on the Dec. arm for the Libya expedition (shown here) are a 1,000 mm f/10 MTO Maksutov, a 500 mm f/5.6 Rubina mirror lens, a Tamron 500 mm f/8 mirror lens and a 300 mm Rubina lens. In the desert the polar aligning was carried out by magnetic compass (allowing for the local magnetic variation) and an inclinometer for the latitude. Image: Chris Baddiley

corona for several minutes. DSLR users shooting at a high altitude Sun suddenly find their camera viewfinder, LCD panel and controls are horrendously placed unless they have brought a 2 m high tripod with them. One neat solution to this neck-breaking problem that I spotted on the Libyan eclipse was that of Ipswich amateur astronomer Mike Harlow (Fig. 11.28). His camera and 500 mm f/8 Tamron telephoto lens were pointing downwards at a gentle angle, pointed at a 10 cm diameter, optically flat mirror which was tilted at about 20° from the horizontal (towards the camera lens). The whole arrangement was mounted atop a short tripod. With the Sun at an altitude of 67°, and the mirror tilted by 20°, the camera could be tilted at a modest 27° (i.e. 67 − (2×20)) down from a horizontal position. An old elliptical Newtonian secondary mirror is ideal for this application especially if the camera is arranged so it points along the longest axis of the mirror for a beefy wide aperture lens. At glancing reflection angles a circular mirror may cut

Fig. 11.28. Mike Harlow's very portable and user-friendly eclipse mount features a large circular flat mirror enabling the camera to be conveniently angled while photographing the reflection of the Sun. In addition, a hinged filter arrangement, attached to the front of the telephoto lens, enables the solar filter to swing out of the light path in seconds, at second contact, and replaced in seconds, at third contact. Note the use of local rubble as a counterweight, thereby avoiding excess baggage charges! The hole in the nearby wall, is, apparently, coincidental . . .. Image: Mike Harlow

off light to the lens if the mirror diameter is only as big as the lens diameter. However with, say, a 500 mm f/8 lens of 62 mm aperture and a 100 mm diameter mirror there is unlikely to be a problem. Remember when examining the images from such a system that they will be mirror imaged.

Strolling further around the Libya desert in 2006, where there were 800 eclipse chasers just from our trip, I noticed just how obvious the veteran eclipse chasers stuck out from the novices. There were just those little subtle touches that only someone who has been on a previous trip sorts out. For example, the experienced eclipse chaser rarely lugs any German equatorial mount counterweights on a trip, not when baggage allowances are so strict. Most mounts will actually track better if they are slightly overloaded and most eclipse chasers have more than one camera. A second camera, a video camera, a telescope, or local rubble can be used as a counterweight rather than a dead mass of useless lead. Many UK eclipse chasers are not impressed by the rip-off prices charged for items like dew caps or filter holders either. Konrad Malin-Smith uses old prune tins to hold his solar filters in place (Fig. 11.29).

Undoubtedly Takahashi's tiny Teegul Sky Patrol II mounting is the answer to the traveling astronomer's prayers. At around $750 it is not cheap, but it outperforms much larger mountings and is very cleverly designed so that the weight of the drive gears themselves acts as a counterweight – neat! Takahashi do not recommend mounting any camera or telescope heavier than 2.5 kg on the Sky Patrol II mount (a 0.65 kg counterweight is provided) but telescopes as heavy as 5 kg, balanced with a similar counterweight, have been carried on this tiny mounting. One minor disadvantage is that the mounting is mainly designed to carry Takahashi telescope adaptors, but it is fairly painless to make a simple adaptor for a camera, as I did for mine: a right angled plate, two holes and two bolts is all that is required.

Fig. 11.29. Konrad Malin-Smith with a unique, light-weight, wooden, collapsible mounting at the 2001 Total Solar Eclipse in Zimbabwe. The lens cap filter holders for his two telephoto lenses and camcorder are made from prune tins! Photographed by the author

On the eclipse trips I have been on I have not come across many organised Japanese eclipse groups (except India '95 where they tried to have us shot!) but a favourite combination with Japanese eclipse chasers is the Pentax 75 mm SDHF 500 mm f/6.7 refractor attached to the Takahashi Sky Patrol II.

Finally in this section on making life easier, you simply do not have to carry huge amounts of kit to an eclipse site to get some great results. By taking lots of pictures, keeping exposure times short and applying expert and patient image processing to the results on your return from the trip, great images can result from modest equipment. Pete Lawrence of Selsey, UK, a neighbour of the great Sir Patrick Moore, obtained many fine images of the 2006 TSE with just an 80 mm short focus refractor, a Canon 10D DSLR and a lightweight tripod (see Fig. 11.30).

Heliostats, Siderostats and Coelostats

Arguably the ultimate observing system for any solar observer is a Heliostat or one of its variants. Put simply it uses an optically flat mirror mounted within or parallel to the polar axis that tracks the Sun; in this way an image of the Sun can be reflected up the polar axis to a fixed position (where a lens or a fixed mirror is placed). Typically it looks, at first glance, like the telescope is staring at the ground. Such an arrangement can just be glimpsed in the figure of Glenn Schneider (12.2) in Chap. 12; with this arrangement the observer/camera/detector does not need to move. The arrangement has to be carefully designed such that the mounting is

Fig. 11.30. Pete Lawrence obtained many fine images of the 2006 TSE using the equipment shown here, i.e. a Skywatcher 80ED Pro 80 mm f/7.5 refractor (reduced to f/5 and approx. 400 mm focal length with a 0.63× telecompressor) a Canon 10D and a lightweight tripod. Image: Martin Andrews

appropriate for the observer's latitude (i.e. in this case the latitude at the eclipse site) and the mirror is large enough to pass the light into the camera lens without vignetting. As small apertures are sufficient for solar observing and eclipse observing the flat mirror does not have to be too large, which is the financial barrier if such an arrangement is used for observing fainter night-time objects. In the Siderostat variant the tracking mirror is not equatorially mounted, so does not have to be re-designed for the latitude of each eclipse site. However, the mirror needs to be tracked in two axes, not one. More often than not the beam of light is fired horizontally to the instrumentation with the Siderostat design. The final variant, the Coelostat, uses two flat mirrors to track the Sun and the final beam is usually directed horizontally or vertically. The advantage of the Coelostat is that the image of the Sun does not rotate around the centre of the field. In the two other designs, while the solar image will stay in the centre of the field (assuming the design and the drive is of high quality) the Sun and the corona will rotate slightly over a few minutes. Of course, with modern software images can be rotated to any position angle you desire, if there are noticeable features to align on, after the pictures have been taken. The individual corona exposures of a second and less will not be affected.

In practice the designs I have seen on eclipse sites have been simple Heliostats designed so that the camera and lens, or the telescope, can be pointed down during totality and not up. This avoids the horrendous neck strain which often arises when you try to see your camera LCD, viewfinder and controls when the Sun is high in the sky. Of course simpler solutions do exist and a 90° clip on magnifier, which attaches to the camera viewfinder can be a life-saver, as can a simple star diagonal for visual viewing. A home-made Heliostat, if designed carefully can be a much more portable and lightweight option than a bulky German Equatorial Mounting too. Large Heliostat systems are generally preferred by ultra-keen amateurs or professionals who have heavy instrumentation to attach to the light gathering end. The amount of power needed to rotate a lightweight mirror to track the Sun is also much less than might be required for a bulky mounting carrying a heavy payload. As always where single mirrors are involved you need to remember that the image you capture is mirrored before you send your final masterpiece out to your friends and to magazines and professional bodies!

Daytime Polar Alignment

Most keen amateur astronomers will be more than familiar with polar aligning an equatorial mounting from their own backyard at night. In the northern hemisphere the bright star Polaris makes life easy; in the southern hemisphere Sigma Octantis can be located in the finder scope and used in the same way. However, how do you polar align a telescope at an unfamiliar site in broad daylight? On every eclipse trip someone will unpack their equatorial mounting on the big day and then realise that they do not have a clue how to polar align it. However, the good news is that as totality only lasts a matter of minutes, and eclipse exposures are short, the only requirement is that the Sun and corona do not drift out of the field. This is far less of an alignment problem than the critical polar alignment required for long Deep Sky exposures; indeed, pointing your mounting's polar axis within a few degrees of the pole will be more than adequate. Most observers are happy with a simple alt-az tripod head. If you are taking an equatorial head, then before your eclipse trip you need to acquire a number of items which will help you polar align your camera/telescope mounting. A quality compass (to find north) is a must but you also need to know the current difference between true north and magnetic north for your eclipse site. A high quality modern map will show this "magnetic declination". You also need a bubble level or a compact pocket spirit level to ensure the base of your telescope's equatorial mount is horizontal. Many equatorial mountings feature a declination circle which can be especially useful for checking your polar axis angle. For example, if you know your latitude on the Earth and you set the declination circle to that latitude, with the mount's declination axis horizontal (e.g. as indicated by the mounting's meridian marker), the telescope tube should point vertically. If it does not you may have mis-adjusted the polar axis angle. A vertical tube can be verified with your pocket spirit level and used to level the mount precisely. Of course, if you are at a hotel for a night or two before the big day you can carry out a nighttime polar alignment, if skies are clear. A double check that you have polar aligned correctly can be carried out by checking that when you get the filtered Sun centred in the telescope the declination circle gives the Sun's declination correctly. If the eclipse occurs after midday at your site then you may have the luxury of being able to observe the Sun as it transits the local meridian, thus giving you total confidence of where the north–south line is. Some caution is required here though. You need a watch which can give you G.M.T. accurately and, before you head for the eclipse you need to make sure you are 100% confident of the time that the Sun will transit from your intended location on the day. Bear in mind that this transit time is

NOT the same as local noon! The Earth's passage around the Sun is an ellipse, not a circle, and it transits due south roughly 14 min late in mid-February and up to 16 min early in October/November. Also, if you are in India their time zone is GMT + 5 h 30 min, i.e. not an even number of hours! Nevertheless, if you have the eclipse site transit time to hand, for your location, it can make the azimuth alignment of your mount very easy. A colleague holding a weight on a string at local noon can create a shadow which you can scratch a line on the ground parallel too at the critical time.

Chapter 12

Some *Very* Keen Eclipse Chasers

In any field of astronomy you will find diehard self-confessed addicts who are obsessed with their particular specialised interest. In the UK supernova patrollers Tom Boles and Mark Armstrong have checked two-thirds of a million galaxy images between them, *by eye*, over a 9 year period, to discover over 170 supernovae. Their fellow countryman Gary Poyner has made over 170,000 magnitude estimates of faint variable stars in a 30 year visual onslaught of his light polluted Birmingham skies. Throughout the second half of the twentieth century George Alcock discovered five comets and five novae using nothing more than binoculars and his phenomenal memory of the night sky: he had memorised the positions and brightness of 30,000 stars! Diehard total eclipse chasers are often referred to as umbraphiles, or even coronaphiles. They want to spend as much time under the Moon's shadow as possible and see as many different total solar eclipses as possible. Of course, since the 1970s, the advent of cheap air travel has made the target of chasing down every total solar eclipse (on average there is a TSE or hybrid somewhere on Earth every 16 months) much more affordable, terrorism notwithstanding. Nevertheless spending several thousand dollars every year or two, scurrying around the world to be under the Moon's shadow, is quite a financial undertaking. It is not just a financial issue either as in most cases family and job commitments will get in the way at some point. Some of the keenest eclipse chasers manage to work around the time and financial restrictions though, and they are not all multi-millionaires! Many of the keenest eclipse chasers take up temporary posts as lecturers or tour guides with the eclipse expedition travel companies. Others are fortunate enough to be associated with university departments or observatories that will fund their trip on scientific grounds. Some eclipse chasers are so obsessively dedicated that they travel to the track of totality one year before the event to check nothing has been left to chance – Amazing! Age does not seem to be a barrier for the toughest eclipse chasers either. Dame Kathleen Ollerenshaw of Manchester travelled on the 2006 trip into the Libyan Desert aged 93! Eric Strach secured his excellent shadow band video in 1998, aged 83. For more normal mortals, if one assumes an eclipse-chasing career could last from age 20 to age 80, i.e. a 60-year span, then in theory one could expect to travel to 45 totalities at most. In practice, the world's keenest eclipse chasers in the early, cheap air travel, twenty-first century, at least those that advertise the fact, are all now sitting on impressive totality experiences in the low to mid-twenties. At the start of the 1980s totality expedition numbers in the mid teens would put you into an elite group and the number of TSE's seen by Professors Charles Smiley (1903–1977),

Some Very Keen Eclipse Chasers

175

Max Waldmeier (1912–2000) and Donald Menzel (1901–1976) seemed almost impossible to overhaul, unless you were associated with a University department specialising in solar studies. In 1999, Frenchman Bernard Milet (then 74) was reported in a few European newspapers as having notched up 23 totalities with the eclipse of 11 August of that year, and the tabloid press declared that this was a world record. However, in recent years all these records are being overhauled by a small group of US eclipse chasers who are young enough to enjoy many more times under the umbra in the twenty-first century.

Keenest of the Keen

At the very top of the current list (unless there are some more travelled reclusive eclipse chasers lurking in the Moon's shadow) are the Umbraphiles Jay Pasachoff, Glenn Schneider and John Beattie of New York City (see Figs. 12.1–12.3) who, after the 2006 total solar eclipse, had all been under the umbral shadow 26 times, and are still keen to see more. When it comes to assessing how many eclipses were actually seen in a clear sky and how many were clouded out, as well as which annular totals were more total than annular at the observer's position, well, then it becomes a greyer area. If a total eclipse is clouded out for all but a split second, and then just glimpsed through thinner cloud, was that seen or not? Either way, if you go by eclipse expeditions that factor does not enter into the equation. Since his first eclipse in 1959, in an aircraft over Marblehead Massachusetts, Jay Pasachoff has been on expeditions to 44 total, annular and partial eclipses (including the annular of September 2006). Twenty-six of these were total eclipse expeditions

Fig. 12.1. Professor Jay Pasachoff, of Williams College, Massachusetts, is one of the "Big Four" eclipse chasers mentioned in this book (i.e. the other three being messrs Beattie, Schneider and Small). Jay is a dedicated umbraphile, having only been completely clouded out once in 26 total solar eclipse expeditions, starting in 1959 October and up to and including 2006 March 29. He is shown here projecting the solar image at the 2005 Annular Eclipse. Image: Jay Pasachoff

Fig. 12.2. Another of the "Big Four" Umbraphiles, Dr. Glenn Schneider of Steward Observatory, University of Arizona, has also travelled to 26 total solar eclipses, with only three being cloudy (between 7 March 1970 and 29 March 2006). The picture shows Glenn at the 29 March 2006 total solar eclipse in Side, Turkey using his "Lug-a-scope", a 100 mm aperture f/12 achromatic refractor fed by a 146 mm 1/8 wave coelostat (solar tracking mirror) sitting on the ground. A Pentax camera is positioned at the focus. Image by kind permission of Glenn Schneider and Jay Friedland

Fig. 12.3. Two living legends of the eclipse chasing world, at Jalu in the Libyan Desert, on 29 March 2006: John Beattie, of New York (*left*, with cap) and Germany's Friedhelm (Freddy) Dorst. Like Pasachoff and Schneider, Beattie has travelled to 26 totalities up to and including 2006. Freddy has travelled to 24, with only three being clouded out. Image: Mike Gill

(up to and including 29 March 2006) and only one was completely clouded out (Hawaii in 1991). Professor Pasachoff is Director of Hopkins Observatory and works closely with Drs Steven Souza and Bryce Babcock. They study the million degree temperatures in the solar corona and their work is supported by the National Geographic Society, the National Science Foundation and the Keck

Astronomical Union's (IAU) working group on solar eclipses and of a public information group when eclipses actually occur.

Glenn Schneider has seen 23 totalities out of his 26 and amassed an impressive cumulative time under the umbral shadow of 1 h 13 min and 38 s, up to and including the 29 March 2006 total solar eclipse. Dr. Schneider is an associate astronomer and the Hubble Space Telescope NICMOS (Near Infrared Camera and Multi-Object Spectrometer) scientist at the University of Arizona's Steward Observatory. Like many addicted umbraphiles with totality totals in the twenties, Glenn Schneider's passion for eclipse chasing was fired on 7 March 1970 when the Moon's shadow tracked up the Eastern coastline of the USA. Schneider was only 15 years old when he witnessed this event but, as he commented on the Solar Eclipse Mailing List (SEML) in 2006: "I'll always remember that first awe-inspiring totality with particular reverence as indeed it crystalised my life's direction". When total solar eclipses cross over or close to North American territory it does have the effect of spawning a whole new breed of umbraphiles. The 7 March 1970 eclipse (maximum duration 3 min 28 s) was one such event and seems to have inspired most of the leading eclipse veterans who have travelled to 20 or more totalities. A similarly inspirational event was the 26 February 1979 eclipse which crossed the northernmost states of the USA before moving into Canada where 2 min 49 s of totality could be experienced. Twelve years later, the 11 July 1991 eclipse track passed over Hawaii, Baja California and Mexico. Not only was this an eclipse which was easy for US citizens to get to, it offered a massive 6 min and 53 seconds of totality. It was an event that inspired this author and thousands of others, although many of us had a very miserable and cloudy experience in Hawaii. Nevertheless, after I had recovered from the huge disappointment of 1991 (it took 3 years) the lure of the corona just inspired me to try again.

John Beattie is the third member of this elite group, all of whom have been under the umbra 26 times. Although John is tied with Jay Pasachoff and Glenn Schneider for the most number of total eclipse expeditions, on 29 March 2006 he briefly held the outright record: he was in Libya and so witnessed totality minutes before Jay and Glenn, who were across the Mediterranean at Kastellorizo, Greece. John Beattie's commitment to total eclipse chasing is 100%. Nothing is left to chance and every alternative option is considered, at every stage. The cruise ship he was booked on for the 1984 eclipse broke down leaving passengers unable to reach the track of totality. At the last minute he got a flight to Noumea in New Caledonia where another chartered vessel was booked to take three dozen eclipse chasers to the track. Many amateurs will recall a cloudy experience (especially in Finland) for the 22 July 1990 eclipse. John Beattie was at Cherskiy in Russia with over 100 other eclipse chasers. However, when the rain started to fall he decided moving vertically was the only option. A number of aircrafts were rapidly commandeered and Beattie, along with a few dozen others witnessed totality from above the clouds. Eleven years later, on the morning of the African eclipse of 21 June 2001 he even tried to get word to the President of Zambia to use his influence to put out a bush fire that might spoil the eclipse observations at Lusaka Airport. Surely John Beattie must be the world's most determined eclipse chaser as well as, jointly, one of the top three umbraphiles on 26 times under the umbra, as of March 2006.

Craig Small, from New Jersey (shown with Mike Gill, the Solar Eclipse Mailing List moderator, in Fig. 12.4), is only just behind the impressive gang of Pasachoff, Schneider and Beattie in terms of totality expeditions having been under the

Fig. 12.4. Mike Gill (*left*) from Birmingham, UK and Craig Small (*right*), from New York, holding Craig's lucky eclipse flag. Mike, the Yahoo Solar Eclipse Mailing List (SEML) moderator has seen 12 out of 14 totalities and Craig has seen all 24 totalities from his 24 total solar eclipse expeditions! Picture taken at Korou, French Guiana, for the September 2006 annular eclipse. Image: Mike Gill

umbra 24 times and seen *all* of his 24 total solar eclipses. Having *seen* 24 totalities though surely makes him one of a re-defined top group, i.e. the big four: Pasachoff, Schneider, Beattie and Small.

Umbraphile Recluses?

Of course, one could argue that there may be some wealthy, reclusive multi-millionaire eclipse chasers who have been to more than 26 totalities and witnessed more total solar eclipses than Messrs. Pasachoff, Schneider, Beattie and Small. Maybe, but it seems unlikely. Firstly, most umbraphiles who have been to more than 20 total eclipses like to broadcast the fact and, secondly, the really tricky eclipse tracks weed out the men from the boys. Michael Gill, the British moderator of the SEML has looked into this aspect in some detail. According to Michael, in e-mail correspondence with this author, the 23 November 2003 TSE over Antarctica would definitely have been witnessed by less than 1,000 people and probably only 800. This number is made up from three umbraphile expeditions (a LanChile flight, a Quantas flight and the icebreaker ship the Kaptain Klebnikov) containing less than 500 eclipse chasers, as well as 300 staff at various Antarctic stations/projects, namely Novolazarevskaya, Mirni, Maitri and Dome Fuji. According to Mike, Jay Pasachoff distributed a questionnaire on that Antarctic Qantas flight asking if there were any eclipse chasers with more TSE's than himself, John Beattie, Glenn Schneider and Craig Small but no-one came forward.

Incidentally, the cost of a cabin on the icebreaker ranged between $18,000 and $35,000, and a seat on the Quantas flight cost between $2,500 and $11,000. Not one for the poverty-stricken eclipse chaser! The Quantas flight was the longest Australian domestic flight on record. Sydney and back to Sydney in just over 14 h (an Antarctic landing was not possible).

The 8 April 2005 hybrid eclipse was another one for the dedicated eclipse chaser. It was total for much of its length, but only over the eastern Pacific Ocean. Mike Gill estimates that the three eclipse chasing ships that were under the track (M/S Paul Gauguin, M/V Discovery and M/V Galapagos Legend) contained, at most 1,500 passengers and crew. There were no reclusive billionaire eclipse-chasing yachts spotted in the area.

However, although Pasachoff, Schneider and Beattie lead the field they cannot afford to relax as there are plenty of other eclipse chasers who are only a few TSE's behind them. Ernie Piini of Santa Clara County, California is one such active umbraphile who has been on 26 eclipse expeditions (including annulars) and is yet another eclipse chaser whose interest was triggered by the 1970 TSE which tracked across Mexico, the Gulf of Mexico and the Eastern USA seaboard. Retired Lt. Colonel Bill Speare, Director of the Spitz Planetarium in Scranton, Pennsylvania, is another top umbraphile and a veteran of 23 solar eclipse expeditions. In Japan the nearest challengers appear to be Osamu & Kazuko Ohgoe who between 1974 and September 2006 had been on 33 solar eclipse expeditions comprising 18 totals, 2 hybrids and 13 annulars.

Germany's top eclipse chaser Friedhelm (Freddy) Dorst also has an impressive tally of 21 successes from 24 total eclipses; he is shown in Fig. 12.3, with John Beattie. Like John, he has a legendary reputation amongst eclipse chasers, especially German eclipse chasers. Freddy is renowned for recording the corona even at *annular* eclipses, by using frighteningly low value (ND3 and ND4) neutral density filters to hide the blazing solar annulus. In addition, the verb *Dorsting*, or "to *Dorst*" an eclipse has gone down in the eclipse chasers dictionary. Freddy is not a tourist. He arrives at the eclipse track, sees the eclipse, and heads straight back to Germany after totality. According to British umbraphile Mike Gill to truly "Dorst" an eclipse you have to see one on a different continent and then be back at work without taking any time off and without your colleagues knowing that you've been away! Cool! In addition, on airborne eclipse flights, such as the 23 November 2003 Antarctica flight, Freddy always packs the same house brick in his onboard hand luggage as he has found that it acts as a good, heavy stabiliser for his camera when photographing totality through a small aircraft window. Seeing a passenger strolling along the aisle with a brick in his hand may be a bit scary in this post 911 era, but it is all part of the experience if you are an eclipse chaser (see Fig. 12.5).

Just below the most fanatical group of eclipse chasers in the "20 + TSE's seen" category lie many more normal mortals who simply try to get to as many eclipses as possible when work, finances and family commitments allow. Many of these have backgrounds in science, engineering or astronomy and are sponsored in some way to travel on these trips. Many more are just keen amateur astronomers with enough time and money to indulge in eclipse chasing.

The most prominent member of this group, with 16 out of 20 totalities seen (up to and including 2006) is the man known, throughout the astronomical world as "Mr. Eclipse", NASA's Fred Espenak (Fig. 12.6). Espenak is an astrophysicist at NASA's Goddard Space Flight Center in Greenbelt, Maryland. However he is best known for his meticulous eclipse prediction work, specifically his two NASA

Fig. 12.5. Not a terrorist, but Germany's leading eclipse chaser Freddy Dorst onboard the 14 h 23 November 2003 Qantas Antarctic eclipse flight. Freddy is discussing the merits of using a brick for keeping a camera steady, when photographing through an aircraft window, with umbraphile Paul Maley. Image by Paul's wife Lynn Palmer and by kind permission of Paul Maley

Fig. 12.6. The man known as "Mr Eclipse", NASA's Fred Espenak, who provides the most comprehensive eclipse track predictions for the world, as well as being a dedicated eclipse chaser himself (16 totalities seen from 20 trips between 1970 and 2006). Shown here in 1999. Image: Patricia Totten Espenak

Fig. 12.7. Prof. Miloslav Druckmüller of Brno University of Technology in the Czech Republic, mentioned throughout Chap. 11, is the undisputed king of eclipse image processing, whether processing his own images or those of other umbraphiles. Image: Miloslav Druckmüller

eclipse catalogues (Fifty Year Canon of Solar Eclipses: 1986–2035 and Fifty Year Canon of Lunar Eclipses: 1986–2035). He also manages NASA's eclipse pages at: http://SunEarth.gsfc.nasa.gov/eclipse/eclipse.html as well as his own highly comprehensive "Mr. Eclipse" pages at http://www.mreclipse.com/. Espenak has also collaborated with *Spears Astronomy Travel* on a number of eclipse expeditions.

Another name that, for decades, has been associated with total solar eclipse photography is Wendy Carlos, a renowned electronic music composer, a veteran of some 20 eclipse expeditions and an expert at coaxing the maximum amount of detail from a photographic negative of the corona. She saw her first eclipse on 20 July 1963, through a cloud gap as the track passed through Maine, yet missed the 1970 totality due to an "airport screw-up". But after that her photographs of totality would amaze and delight the world. As we saw in Chap. 11, when the digital era dawned a new king of corona imaging was borne. The power of digital scanners and digital cameras was fully harnessed by the Czech mathematician and eclipse chaser Miloslav Druckmüller, shown in Fig. 12.7.

Another European umbraphile more than worthy of a mention is the Belgian amateur astronomer Patrick Poitevin. A veteran of 34 solar eclipse expeditions from 1976 to 2006 Patrick has travelled to 15 total solar eclipses, one hybrid (annular-total) eight annulars and ten partials with only one of his totalities (Finland 1990) being cloudy. Patrick started the Solar Eclipse Mailing List (SEML) in December 1997. In July 2004 it moved to *Yahoo! Groups* (http://tech.groups. yahoo.com/group/SEML/) where it is supervised by British eclipse chaser Michael Gill, himself a veteran of 14 total solar eclipse trips, with only two being clouded out.

British Eclipse Chasers

I could not leave this Chapter without a mention of some of the keenest British eclipse chasers and eclipse characters, many of whom I have travelled on eclipse trips with. Personally, I would not travel to some of the godforsaken poverty-stricken third world dumps I have been to without their company. In the UK, astronomer Patrick Moore is the face the public most associate with any form of astronomy. His monthly television program *The Sky at Night* celebrates 50 years on the BBC in April 2007; a staggering achievement. Patrick's first total solar eclipse was the June 1954 event which he observed from Sweden. He followed this with a ground-breaking TV "first" with live commentaries from various parts of Europe for the 1961 eclipse; Patrick himself was on Mount Jastrebač in Yugoslavia for that one. After that he travelled to Siberia (1968), off the coast of Mauritania on the good ship Monte Umbe (1973), the East Indies (1983), Bali (1986), the Phillipines (1988), Baja California (1991), Peru (1994), the China Seas (1995), near Aruba in the Carribean (1998) and Cornwall, England (1999). So, Patrick has been on twelve total eclipse expeditions but was only under total cloud in his own country in 1999. For the 1973 trip on the good ship Monte Umbe Patrick travelled with *Transolar Tours*, a Merseyside company which offered eclipse and astronomy related holidays. From 1981 the astronomy holidays offered by *Transolar Tours* moved from David McGee to brother Brian McGee as a new company *Explorers Tours, Explorers Travel Club* or just *Explorers*. Under Brian McGee the business has now been offering Eclipse holidays for over 25 years and has a regular flock of several hundred eclipse chasers for every solar eclipse trip they organise. They are also now the UK's biggest Scuba Diving holiday company! For the 1983, 1986, 1988, 1991 and 1994 eclipse trips Patrick was the *Explorers* eclipse lecturer, ably assisted by Dr. John Mason who still continues his inimitable eclipse lectures with *Explorers* to this day. Indeed, most travellers with *Explorers* regard John's pre- and post eclipse round ups as a highlight of the trip, second only to the eclipse itself! The Libya 2006 Totality was the 12th time under the umbral shadow for John and a fundamental item he carries to each eclipse expedition is his eclipse hat, the magical properties of which can easily part the clouds as second contact approaches!

Perhaps the UK's most experienced eclipse chaser is Mike Maunder (Fig. 12.8). Mike is an expert on film based photography and his company, *Speedibrews*, supplied specialist developing chemicals to the amateur community for decades. Since the June 1973 "Eclipse of the Century" totality he has been a relentless eclipse chaser. To date (2007) he has travelled to 18 totalities and six annulars, although he did not have to travel far for the 1999 totality, just to a good spot near his Alderney home. He also travelled abroad for the 1999 Transit of Mercury, the 2004 Transit of Venus and the occultation of Sigma Sagittarii by Venus in Kenya in 1981. Mike is also the joint author of two relevant books with Patrick Moore, namely "The Sun in Eclipse" and "Transit", both published by Springer.

My good friend Nigel Evans (see Chap. 8, Fig. 8.1) is one of the UK's keenest globe-trotting astrophotographers and has been travelling to exotic locations, for astronomy, and just for sightseeing, his entire adult life. As well as being an eclipse chaser, Nigel witnessed the launch of Apollo 17 on his first foreign astro-holiday, as an 18 year old, and successfully captured the Leonid meteor fireballs of 1998 and the storms of 1999 and 2001 with his equipment. He has now travelled to 12 total solar eclipses, and one annular, with only the Finland (1990) and Mongolia (1997) totalities being clouded out. Nigel's approach is simple: take as much camera equipment as possible to the eclipse and do everything! After third contact he can always be heard cheerfully admitting "Well, that was another total disaster" and yet, weeks later he always salvages some impressive results from the day. Travelling on international flights from the UK means Nigel and his wife Alex are

Fig. 12.8. Mike Maunder, one of the UK's keenest eclipse chasers, and a veteran of 18 total solar eclipses and six annular eclipses. Photograph by kind permission of Val and Andrew White

each invariably limited to baggage limits of 20 kg plus 5 kg of hand luggage. It goes without saying that this is usually stretched to almost 30 and 10 kg in practice, with an emergency procedure of unloading items into separate carrier bags if the check-in personnel get tough! The second fall-back procedure is unloading heavy lenses onto friends whose baggage is nearer the proper limit! It goes without saying that the vast bulk of the 80 kg (2 people × (30 + 10)) payload carried by Nigel and Alex on each eclipse trip is photographic. Clothing is carried mainly as packing material for delicate cameras and lenses. Nigel carries an equatorial mounting along with 500 mm, 1,000 mm, fish-eye and flash spectrum lenses to each eclipse. Then there is his multi-Sun camera and the *Psion* controller that controls (sort of!) the automated photography, along with the wires, cables and GPS unit. At the end of each eclipse trip Nigel and Alex can be found in their hotel room or cabin, sitting on their case lids and trying to cram all Nigel's cameras and tripods back in for the journey back!

On my 1994, 1995 and 1998 eclipse trips the unmistakable figure of the rock guitarist Brian May (Fig. 12.9), from the rock band Queen, was on the same holiday package as me, travelling with the British travel company *Explorers*. Many people are unaware that Brian studied Physics at Imperial College and is a keen amateur astronomer and friend of Patrick Moore. Brian kindly agreed to being interviewed by me, on my 1998 holiday video, about his love of total solar eclipses. He revealed that his best eclipse was the big 7 min one of 1991 where he went to San Jose del Cabo in Baja, California; he had actually heard about it happening just by chance and decided to go there. But as far as his most memorable eclipse was concerned he thought he would never forget the cloudy *Explorers* 1997 Mongolia eclipse where they were chasing about on buses in a snow storm up to totality which was "a good little wheeze". You meet some fascinating people on eclipse holidays!

World renowned Space and Sci Fi artist David Hardy (Fig. 12.10) is another well-known UK eclipse chaser and, unlike the vast majority of Umbraphiles, his finished product after each eclipse is a painting, and not a photograph or a digital image. David illustrated Patrick Moore's book Suns, Myths and Men, in 1954, when he (David) was only 18. He had 5 days to complete the eight illustrations before he started his National Service in the RAF. Two of David's paintings are reproduced here, both depicting the 2006 eclipse in Libya. David is also the author of a science fiction novel called *Aurora*.

Sheridan Williams is also very well known in UK eclipse chasing circles and amongst eclipse chasers worldwide. His website (http://www.clocktower. demon.co.uk/eclipse.htm) contains some very useful data on eclipses and a lot of statistics relating to individual observers. His book *UK Solar Eclipses from year 1 to 3000 AD* was published prior to the August 1999 UK eclipse and is a unique database of British and Irish solar eclipses. Sheridan has appeared many times in

Fig. 12.9. The author (*right*) with rock star, astronomer and eclipse chaser Brian May (Queen guitarist) at the Peru/Chile border in November 1994. Photograph by the author's father, Denys Mobberley

the UK media in recent years whenever a major eclipse is due and he has acted as an eclipse tour guide for Ancient World Tours (http://www.ancient.co.uk/).

Professor John Parkinson of Sheffield Hallam University is another well-known UK eclipse chaser who often appears in the British media before, during and after total solar eclipses. He is easily recognised by his brightly coloured waistcoats and eye-catching T-shirts! Like another "J.P.", Jay Pasachoff, he has a professional interest in learning more about the extreme temperatures in the solar corona.

Fig. 12.10. The renowned British space artist and eclipse chaser David Hardy who painted Figs. 10.1 and 10.2. Image: David Hardy. http://www.hardyart.demon.co.uk/

Chapter 13

The H-Alpha Revolution

H-Alpha Viewing

As we saw in Chap. 2, solar prominences were, not surprisingly, first described in observations of total solar eclipses. The very first detailed descriptions were probably by Vassenius in 1733, although Stannyan, in 1706, may have described them too. After the eclipse of 19 August 1868, Jules Janssen in France, and Norman Lockyer in England devised methods of observing solar prominences spectroscopically. Charles Young obtained the first photograph showing a prominence at the total solar eclipse of 1870. In 1924 George Ellery Hale invented the spectrohelioscope, which essentially uses a spectroscope to scan across the solar surface and produce an image at any desired wavelength. The spectrohelioscope is a complex beast involving moving slits or mirrors, such that the narrow spectroscopic slit effectively scans across the solar surface. However, a few are even in amateur hands. An excellent book entitled *The Spectrohelioscope* has been written by Fredrick Veio. In the UK the well-known amateur astronomers Cdr Henry Hatfield and Brian Manning are perhaps the best-known owners of such fascinating machines. Such complex machinery is rather off-putting for most amateur astronomers, who traditionally, have been content with just observing sunspots by projecting the image or using safe solar filters, but fortunately huge strides were made in producing narrow band solar filters for amateurs throughout the 1970s, 1980s and 1990s. In recent years there has been an upsurge in amateur astronomers who are imaging the Sun at H-Alpha wavelengths, i.e. at 656.28 nm. At this wavelength the limb prominences can easily be seen, even while the usually lethally bright Sun is in the field. More expensive, ultra-narrow band, H-Alpha filters can reveal H-Alpha features on the disc too. Normally it takes a total solar eclipse to reveal these features. However, if you think these filters are a substitute for seeing a total solar eclipse, think again! Nothing can create the awe and spectacle of the shadow of the Moon passing over you, the black disc of the Moon sliding slowly into place and the subtle and stunning sight of the solar corona, let alone the surreal effects like shadow bands. You do *not* get a spiritual experience from using an H-Alpha filter! However, H-Alpha telescopes are very popular at solar eclipses, especially just before totality, when eclipse chasers want to know if any good prominences might come into view. One advantage an H-Alpha filter does have is that you can see the entire solar limb simultaneously, which you cannot at a total solar eclipse (except at short totalities when the Moon is barely larger than the Sun).

Complex Technology

It should perhaps be explained here that the problem is not simply one of standard filtering of colours. The bandwidth of a good H-Alpha filter is, typically, less than 1 Å (0.1 nm), i.e. only one three thousandth of the visual spectrum, and the filter production is highly complex and expensive. Decades ago Edwin Hirsch of the Daystar company was the sole supplier of such narrow-band filters for amateurs, but recently the Tucson based company Coronado (http://www.coronadofilters.com) have been at the forefront of this technology and have developed a number of exciting products using advanced laser techniques. On the UK's Isle of Man, Solarscope (http://www.sciencecenter.net/solarscope/doc/about.htm) also offer quality H-Alpha filters of 50 mm aperture. Both of these modern companies produce precisely tuned, ultra narrow line width classical Fabry–Perot air spaced "etalons" for their H-Alpha filters. An etalon consists of a matched pair of ultra fine pitch polished and accurately figured fused silica plates. These have partially reflective and low absorption coatings for the desired transmission wavelength. To guarantee the essential fixed air space, the two etalon plates are skillfully assembled with the use of optically contacted spacers. Such filters have a very high throughput at peak resonance and a very narrow spectral transmission.

As one narrows the filter bandwidth centred on the 656.28 nm H-Alpha line the prominences become more and more sharp, and fine H-Alpha features on the disc emerge too. In the 1980s the Baader company advertised prominence telescopes in which a metal disc could be used with a telescope of a specific focal length to exactly occult the blinding solar disc. Using this method even a wide (and less expensive) 10 Å H-Alpha filter would show the prominences, while the dazzling solar surface was hidden behind the metal disc. However, by moving to expensive, narrower bandwidth filters the prominences and subtle surface chromosphere features can be viewed simultaneously. Coronado make filters and small quality refractors optimised for use with such filters. The 2007 Coronado range consists of H-Alpha telescopes from 40 to 90 mm aperture (ranging from $1,700 to $12,000 in price) as well as individual filters priced from $900. These units typically have bandpasses less than 0.7 Å. By stacking two matched H-Alpha filters together, a bandpass finer than 0.5 Å can result. Recently, Coronado's wider, 1 Å bandpass, 40 mm aperture f/10 PST or Personal Solar Telescope (see Fig. 13.1) has made H-Alpha imaging affordable to many and, coupled with a webcam, spectacular pictures of prominences can now be obtained for an outlay of only $450. The PST is mainly just a prominence telescope, and will show few fine details on the solar disc, but it is a remarkable price breakthrough. It may be thought that 40 mm is a very small aperture, but it is sufficient to resolve prominences only a few arcseconds in width and perfectly compatible with typical daytime seeing. Like nearly all H-Alpha systems a filter "de-tuning" collar is provided to optimise the view and, in use it gives a sort of "3D" effect as tweaking it can enhance major disc detail or limb prominences. The reason for this "de-tuning" is that solar flares and coronal mass ejections (CMEs) are amazingly fast-moving events which can mean they are Doppler shifted to be outside the narrow passband of the filter. De-tuning moves the passband, typically by as much as ±1 Å to enable such fast-moving features not to be missed!

Although Coronado make dedicated and safe H-Alpha telescopes, many dedicated H-Alpha imagers choose not to go down this route. Why? Well, with today's webcams and high speed USB 2.0 imagers a 90 mm aperture is simply not always

Fig. 13.1. The 40 mm aperture Coronado Personal Solar Telescope (PST) that has revolutionised H-Alpha observing. Image: Maurice Gavin.

enough when you get good seeing. Daytime seeing is usually very poor, but, sometimes, it is good enough to allow instruments of 150 mm aperture, and larger, to reach their full potential. In addition, imagers who already own, say, a quality 100 mm refractor, will be loathe to shell out thousands of dollars more than necessary when they already own the optics, but not the filter. (Coronado's 90 mm filters can easily be attached to most of the TeleVue apochromats). Of course, it is VITAL that the solar imager who constructs his own H-Alpha system does it safely. There are dangers in making, or modifying, any solar observing system, unless you never plan to use it visually. In general there are two approaches to the H-Alpha telescope system. It is not solely about manufacturing a high quality etalon. In the first, a front aperture filter, (usually with additional filters at the eye end) fits over the front aperture of the telescope. Thus, Coronado's narrowband Solarmax filters can be fitted to, e.g. a TeleVue refractor and, at the eyepiece end, within the mirror diagonal, another interference filter, namely, a Coronado blocking filter (BF) is used. Neither component can be used alone. The focal length of your system determines which BF filter (diameter in mms) you need to accommodate the whole solar image. Thus, a BF10 is for focal lengths of up to 1,000 mm (solar disc 9 mm max), a BF15 for up to 1,500 mm (solar disc 13 mm max), and a BF30 for up to 3,000 mm (solar disc 26 mm max).

In the second type of approach, e.g. for the observer with a larger aperture and long focal length telescope, who may well want to buy his H-Alpha filters from the Daystar company, the main filter is at the eyepiece end, but an energy rejection filter (ERF) is also required to reduce the invisible radiation levels so eye and filter damage cannot occur. The absolute minimum energy rejection required is roughly 1 part in 100,000 for UV wavelengths; 1 part in 1,000 for near IR wavelengths and 5 parts in 1,000 for the far IR. The ERF does not significantly attenuate the visual band as that is where the H-Alpha line resides. (For the visual part of the spectrum 5 parts in 100,000 is normally considered the borderline safety point, but, for safety, 1 part in 100,000 is usually adopted). Of course, the H-Alpha bandpass of under 1 Å is so narrow anyway, literally thousands of times narrower than the visual spectrum, that no massive filtering at this precise wavelength is required for a comfortable visual view. In practice a high quality 0.7 Å filter, placed in the visual band, will let through about 10% of the incident light within that narrow waveband. Essentially, the ERF on a system where the main H-Alpha filter is at the eyepiece end, is there to make sure the invisible IR and UV wavelengths do not get through to cook the filter and your eye. Recently Baader planetarium has offered a new range of premium ERFs called Cool-ERF filters from 70 to 180 mm diameter. These filters incorporate a sophisticated coating that produces a really cold focus by eliminating the vast majority of the infrared radiation.

Daystar's filters, used by most dedicated amateurs with apertures above 90 mm, come in a variety of designs incorporating a tuning T-Scanner which compensates for temperature variations from the nominal 23°C the filter is optimised for, and a heated version which keeps the filter at that temperature. Daystar filters with passbands as low as 0.3 Å can be acquired, but for optimum performance an f-ratio of f/30 or higher before the filter, is necessary.

In 2006 Coronado introduced another affordable narrowband filter for amateur astronomers: a Calcium filter for viewing the violet lines at (3,933 and 3,967 Å). Such filters show the glowing hydrogen plage regions surrounding sunspots, and the super-granulation especially well.

H-Alpha Imaging

With the plethora of imaging devices now on the market the amateur astronomer is spoiled for choice as to what device he or she can safely use to image the Sun. In addition, even though H-Alpha (and Calcium) filters are perfectly safe, you can have the added confidence that, if anything goes wrong, you will only damage the detector and not your retina. When amateur astronomers first look through an H-Alpha filter they are struck by the deep red colour of the image. It is immediately obvious that this is a redder red than you see in everyday life, and the colour may well be a bit off-putting at first. However, this is no problem for the CCD detector in a webcam which is very sensitive at 656 nm. A monochrome webcam like the ATiK 1HS or the Lumenera SKYnyx 2-0 is the best choice as it is unnecessary to use a colour webcam for such narrow-band work. However, digital SLRs, and even small non-SLR digital cameras have been successfully employed and even hand-held to the eyepiece! The Scopetronix company make excellent digital camera to eyepiece adaptors for this purpose (http://www.scopetronix.com).

As I have already mentioned Coronado's $450 PST has sparked a revolution in H-Alpha solar imaging but the PST itself does have quite a few drawbacks where imaging is concerned. The PST's focal plane only protrudes a tiny distance from the eyepiece end necessitating short Barlow lenses, eyepiece projection, or afocal (looking through the eyepiece with the camera lens) imaging. This focal plane restriction has forced many amateurs to dismantle their PST's; always a *very* risky procedure where the Sun and eye safety is involved, unless, you plan to use the device purely for imaging. Remember, the PST has built in BFs to keep damaging IR and UV radiation out of your eyes. Another potential problem, when imaging the whole Sun, is that it is often the case that features on one limb are optimally tuned, but appear out of focus on the other limb. Such anomalies can be overcome by taking various images and combining different quadrants of the Sun into a mosaic. Some very impressive images have been taken with the humble PST by the British imager Pete Lawrence (see Figs. 13.2–13.4).

Fig. 13.2. An image by Pete Lawrence of prominences on the solar limb, taken on 22 April 2006 with a Coronado PST, an Atik 1HSII webcam and a 3× Barlow. Only the limb prominences were exposed. The overexposed disc has been digitally covered with a black mask.

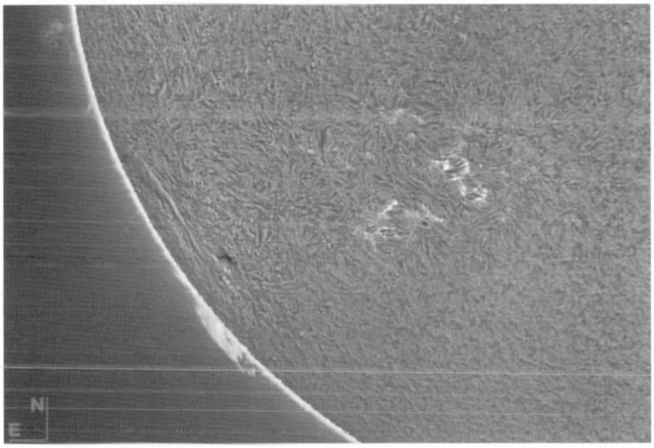

Fig. 13.3. Another Coronado PST image by Pete Lawrence. This is a composite of one image exposed for the prominences and one exposed for the disc. A Lumenera SKYnyx 2-0M and 2.5× Barlow was used on 6 September 2006.

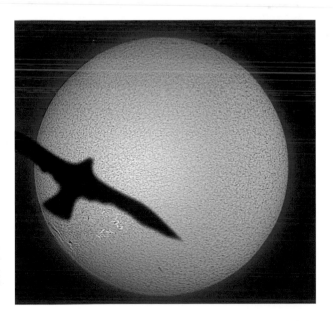

Fig. 13.4. This unusual image shows a bird in silhouette against the solar disc in H-Alpha on 6 September 2006. Lumenera SKYnyx image with a Coronado PST by Pete Lawrence.

One problem often experienced by both solar and lunar observers is that of stray light reflecting off surfaces within the optical train and ruining an otherwise great image. In many cases, where a small area of the Sun is being imaged, this is caused by unwanted filtered light from the *whole* solar disc. With a bit of detective work it is often found that the source of the problem is light from the whole filtered Sun entering the Barlow lens and scattering around the shiny interior of the Barlow itself. The nozzle of the CCD camera/webcam is often wide enough that this scattered light can end up hitting the CCD detector. There are two simple solutions to this problem. Firstly, the webcam/CCD nozzle can be fitted with a cap/diaphragm such that the CCD effectively peers through a hole little bigger than itself, into the Barlow. Secondly, a black card aperture stop can be fitted to the Barlow lens to restrict the area of the Sun being accepted to just that of the region being imaged. Some trial and error experimentation is often needed in this area to determine the optimum aperture stop required.

Some DSLR users have reported an interference fringe pattern effect when using an H-Alpha system, with the very narrow wavelength causing an optical interference between the DSLR camera's inbuilt filter and pixel grid. To avoid such problems you can buy a DSLR from Hutech that has been modified to remove the cameras IR BF, or you can investigate existing users of DSLRs who do not have a problem, to see which DSLR and filter arrangement works for them. The Canon 20Da model, optimised for astrophotography, still features a filter in front of the CMOS detector, but that filter has been modified to let through more of the IR wavelengths than normal. Alternatively you can just use monochrome webcam imaging. Webcams and video cameras have far less pixels than a DSLR so you may wish to take wide field images by imaging loads of small regions and creating a mosaic in *Photoshop* or *Paintshop Pro*. For example, in the latter package the image canvas can be enlarged and the "clone" brush used to copy images onto the larger canvas.

Because high resolution imaging is so seeing dependent using the same techniques as the planetary webcam observer (see my Springer book, *Lunar and Planetary Webcam User's Guide*) will secure the best images, e.g. by using a monochrome webcam and processing the best stacked frames in Cor Berrevoet's *Registax* package. Of course, once you have the final stacked solar image there is nothing to stop you colouring it a more pleasant yellow/gold colour in *Photoshop* or *Paintshop Pro*. The Sun, in H-Alpha is simply a deep red colour with no other colours present. Many imagers prefer to simply represent this as a monochrome image, especially if obtained with a high speed monochrome camera like Lumenera's SKYnyx. However, some artistic license is permitted in this field and it seems to be popular for imagers to convert their images from monochrome to RGB and then reduce the blue channel to zero, raising the red channel value and then tweaking the green channel to tint the whole picture to the right shade of orange or red. Making the prominences fiery red in the best prominence shot and then adding the best globe image detail as a more orangey colour gives the impression of a really angry and exciting Sun, rather different to the weird deep red view seen visually through the same instrument. Solar observers often find that the limb regions appear far too dark when compared to the central disc (even in white light images) once a bit of contrast has been applied. However this can easily be cured by a bit of tweaking with the gamma function in *Paintshop Pro* or *Photoshop*.

In general, and especially in the cheaper, wider bandwidth, filters, the solar disc is much brighter than the prominences and each may require a different exposure/degree of image processing in, e.g. Registax (if a webcam of high frame rate camera is being used to freeze the seeing). So, for whole disc images with a DSLR, the best solution is to use a longer exposure setting to record perfect prominences, but with an over-exposed disc, and a shorter exposure setting to just expose the disc features correctly. Each image can then be optimally processed and, with your favourite software package, simply crop out the disc detail within the solar diameter from the shorter exposure image and paste it on top of the brighter prominence image. If you are purely interested in limb prominences on a tiny section of the limb, the overexposed disc can be digitally masked with a black "fill" command in *Paintshop Pro* or *Photoshop*.

If you are using a digital camera you may well find that different colour channels in the RGB image show different amounts of contrast, even though the image is supposed to be purely red, and the green and blue channels should be black! This is because digital camera CCD or CMOS green and blue pixel filters have some red leakage and, remarkably, the contrast in the green and blue channels can sometimes be more useful than the red (e.g. if there is a moiré patterning artefact)! With apertures of 100 mm or so some truly spectacular images of even modest solar prominences can be obtained. Figure 13.5 shows an image by Damian Peach using a 100 mm Takahashi refractor and a Daystar H-Alpha system, coupled to a Lumenera SKYnyx. Figure 13.6 shows Damian's current solar equipment: a rare 150 mm f/9 aperture Vixen ED refractor with ERF, Daystar H-Alpha filter and his imaging equipment.

Finally, whatever type of solar observing you do, do it safely. If in any doubt, let the camera and the webcam do the imaging and *not* your eye. Retinas cannot be replaced.

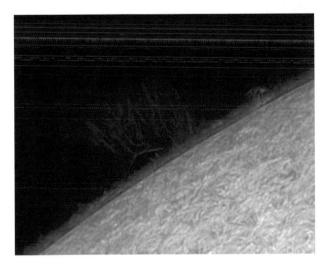

Fig. 13.5. This fine image of a prominence on the solar limb was taken by Damian Peach using a 100 mm Takahashi apochromatic refractor and a Daystar Hydrogen Alpha filter. A Lumenera SKYnyx USB 2.0 camera was used.

Fig. 13.6. A near-optimum equipment set-up for H-Alpha imaging. Damian Peach currently uses a 150 mm f/9 Vixen ED apochromat (a very rare telescope) with a 128 mm energy rejection filter (*top inset*) and a Daystar sub-Angstrom H-Alpha filter. The lower inset shows the Daystar filter in more detail, plus the Lumenera SKYnyx USB 2.0 imaging camera and laptop.

Useful Eclipse Websites, Software and Books

General Eclipse Chaser Sites

"Mr. Eclipse" Fred Espenak's personal web site: http://www.mreclipse.com/
NASA (Fred Espenak) pages: http://SunEarth.gsfc.nasa.gov/eclipse/eclipse.html
Miloslav Drückmuller's eclipse images: http://www.zam.fme.vutbr.cz/~druck/eclipse/
Jay Anderson's eclipse weather/maps site: http://home.cc.umanitoba.ca/~jander/
Jay Pasachoff's Williams College site: http://www.williams.edu/astronomy/eclipse/
Glenn Schneider's site: http://nicmosis.as.arizona.edu:8000/UMBRAPHILLIA.html
Sheridan Williams Eclipse Site: http://www.clock-tower.com/eclipse.htm
Sheridan Williams Totality Stats.: http://www.clocktower.demon.co.uk/total.htm
SEML Yahoo site: http://tech.groups.yahoo.com/group/SEML/
Val and Andrew White's site: http://www.vanda.eclipse.co.uk/
Fred Bruenjes' site: http://www.fredbruenjes.com/
Kryss Katsiavriades' site: http://www.krysstal.com/ecintro.html
Wendy Carlos' eclipse site: http://www.wendycarlos.com/eclipse.html
David Hardy's 2006 page: http://www.hardyart.demon.co.uk/pages-projects/eclipse.html
IAU working group: http://www.williams.edu/astronomy/IAU_eclipses/
IAU eclipse information: http://www.eclipses.info/
2006 Eclipse visualisation (unusual!): http://scalingtheuniverse.com/eclipse/

Travel Companies Specialising in Eclipse Travel

Explorers (UK Eclipse specialists since 1980): http://www.explorers.co.uk/astro/
TravelQuest: http://www.tq-international.com/index.htm
Wilderness Travel: http://www.wildernesstravel.com/
Spears Travel: http://www.spearstravel.com/astronomy/
Ancient World Tours: http://www.ancient.co.uk/
Ring of Fire Expeditions: http://www.eclipsetours.com/

Eclipse Prediction/Camera Software

Heinz Scsibrany's WinEclipse software: http://www.lcm.tuwien.ac.at/scs/welcome.htm
Charles Kleupfel's EclipseComplete software: http://www.zephyrs.com/eclcomp.htm
Charles Kleupfel's Eclipse WinMap software: http://www.zephyrs.com/eclwin.htm
Richard Monk's Eclipse track Pushpins: http://homepage.ntlworld.com/rimonk/index.htm
Chris Venter's DSLR Focus: http://www.dslrfocus.com/
Mike Unsold's Images Plus: http://www.mlunsold.com/
Glenn Schneider's Mac software, Umbraphile: http://balder.prohosting.com/stouch/UMBRAPHILE.html
Steve Bryson's iAstrophoto Mac software: http://homepage.mac.com/stevepur/astrophotography/iAstroPhoto/
High Dynamic Range Software: http://www.hdrsoft.com
Full Dynamic Range software: http://www.fdrtools.com
Registax: http://registax.astronomy.net/
Paul Beskeen's Canon control page: http://www.beskeen.com/astro/SerialDSLRControl/SerialPortControlCables.html
Breeze Systems DSLRRemote Pro: http://www.breezesys.com/DSLRRemotePro/index.htm
Fred Bruenjes camera control software: http://www.moonglow.net/eclipse/photo_software.htm

Camera Adaptors

Scopetronix Digital Camera adaptors: http://www.scopetronix.com

Solar Filters

B. Ralph Chou's page: http://www.mreclipse.com/Special/filters.html
Thousand Oaks: http://www.thousandoaksoptical.com/
Rainbow Symphony: http://www.rainbowsymphony.com/soleclipse.html
Eclipse glasses UK: http://www.eclipseglasses.co.uk/
Coronado: http://www.coronadofilters.com/
Daystar (H Alpha): http://www.daystarfilters.com/hydrogen.htm
Solarscope: http://www.sciencecenter.net/solarscope/doc/about.htm
Baader Planetarium: http://www.baader-planetarium.de/zubehoer/zubsonne/energie.htm

Miscellaneous

SOHO spacecraft: http://sohowww.nascom.nasa.gov/
Measuring the solar diameter: http://www.iota-es.de/soldiam.html
DSLR Camera Reviews: http://www.dpreview.com/reviews/
Heliostats: http://www.observatoryscope.com/heliostat/heliostat.html

Eclipse Paintings

David Hardy's Astro Art: http://www.astroart.org

Five Millennium Canon of Solar Eclipses: −1999 to +3000

NASA Technical Publication TP-2006-214141
by Fred Espenak and Jean Meeus

Main Web Page
http://sunearth.gsfc.nasa.gov/eclipse/SEpubs/5MCSE.html
Online Catalog
http://sunearth.gsfc.nasa.gov/eclipse/SEcat5/catalog.html

Bibliography

Totality: Eclipses of the Sun, by Mark Littmann, Ken Willcox & Fred Espenak. Oxford University Press 1999. ISBN 0-19-513179-7 (paperback).

Fifty Year Canon of Solar Eclipses: 1986–2035 by Fred Espenak. NASA Reference Publication 1178 (Revised) and Sky Publishing 1987.

Eclipse! The What, Where, When, Why & How Guide to Watching Solar & Lunar Eclipses, by Philip S. Harrington. John Wiley & Sons 1997. ISBN 0-471-12795-7.

Eclipses 2005–2017: A Handbook of Solar and Lunar Eclipses and Other Rare Astronomical Events, by Wolfgang Held. Floris Books 2005. ISBN 0-86315-478-6.

Glorious Eclipses: Their Past, Present and Future by Serge Brunier and Jean-Pierre Luminet. Cambridge University Press 2000. ISBN 0-521-791480.

The Sun in Eclipse by Michael Maunder and Sir Patrick Moore. Springer-Verlag 1997. ISBN 3540761462.

The Cambridge Eclipse Photography Guide: How and Where to Observe and Photograph Solar and Lunar eclipses. Cambridge University Press 1993. ISBN 0-521-456517.

UK Solar Eclipses from Year 1, by Sheridan Williams. Clock Tower Press 1996. ISBN 1-85142-093-2.

Jean Meeus: Elements of Solar Eclipses: 1951–2200 Willmann-Bell 1989. (Equations and examples of how to calculate solar eclipses). ISBN 0943396212.

Index

Index

Printed in Singapore